The Discovery of Germs

John Krieger

Published by Lyncean Publishing
Copyright ©2021
ISBN: 978-1-7379676-1-3

CONTENTS

Introduction

Sickness lurks in filth and slime. The stagnant muck of sewers and swamps, or even a bad basement, threatens your well-being and maybe even your life. Living in cleanliness, sunshine, and fresh air is the way to good health. This much has been obvious to most people since primitive times. But what is it about foul places and stinky things that makes our inner functions go haywire? What exactly is the reason for our distress and discomfort? What is the cause of our disease?

Sometimes the cause of a sickness might be obvious — we ate or drank something harmful, or we were bitten by a venomous creature. Sometimes poisonous substances enter our bodies, and that makes us sick. Or maybe we inhaled something we shouldn't have, something that was not fresh air. Disease has been associated with mists and odors and stinky marshes since ancient history. (There's even a disease named "bad air" in Latin: *malaria*.) Maybe some diseases are caused by harmful poisonous fumes swirled into the air we breathe.

But there is another possibility. Maybe there are invisible living things, like bugs, but so small that we can't see them. And maybe they somehow get inside us and cause malfunctions in our internal machinery. Instead of invisible poisons hidden in our food or our air, maybe there are invisible living pests and parasites hidden in the environment around us. This living parasite explanation works especially well for plagues and epidemics. Sometimes there is a malady that spreads rapidly through a community, and we can explain this by saying that there are unseen living bugs passing from person to person. Maybe a plague is caused by tiny living creatures sweeping through communities of people the way swarms of locusts sweep across the land.

Even 1500 years before the invention of the microscope, this was not a particularly difficult guess to make. Sometimes we can observe infestations of worms living inside livestock. Sometimes we find that we have an itchy bug bite, but we never saw the bug. If there are harmful creatures barely large enough to see, maybe there are also harmful creatures even smaller than that.

But how could things so tiny that we can't even see them cause so much devastation? Mosquito bites are usually less bothersome than tiger bites. Being hit by a grain of sand is less damaging than being hit by a rock. What about creatures even smaller than a speck of dust? How could they be powerful enough to harm us? If our eyes can't see them when we look right at them, and our skin can't feel them when we touch them, how could they cause any damage to our lungs or our guts? Animals get sick, too, even animals much larger than we are. How could something we can't even see be powerful enough to bring down a huge cow?

Not all health problems are caused by an outside agent attacking us. Some problems seem to be associated with one's diet or temperament, and ancient scholars talked about an internal imbalance of "humors" as a cause of disease. (Today, we might talk about vitamin deficiencies or hormonal imbalances.) But some health problems seem to be due to a hostile outside factor. Since ancient times, people with scientific minds have looked at human disease and recognized that there is something out there. Somewhere out there is an unseen agent of disease, spread around the world and threatening us from its hiding place. It often seems to reside in rotting things and smelly places, and we naturally try to avoid these things. But what is this invisible dangerous something? The scholars suggested different answers. Some guessed that this agent of disease was like an airborne poison, and they talked about "miasmas." ("Miasma" comes from an ancient Greek word meaning "defilement" or "pollution.") But there were a few others, scattered through Ancient Rome, Medieval Arabia, and Renaissance Europe, who guessed that this unseen thing was alive. Or at least as alive as seeds are. Instead of invisible poisons, some people guessed that there were invisible seeds floating around, like pollen perhaps, and when these seeds enter someone, they sprout or "germinate" into disease. These seeds are the "germs" of disease.

But nobody knew for sure. All anyone had was guesses, and all the scholars could do was bicker with each other. Whatever this "something" is, nobody could see it, nobody could watch how it worked, and nobody could do any experiments with it. Nobody could show, one way or another, what really caused disease. Without plenty of evidence and careful study, without *scientific proof*, there was no knowledge, and no cure.

Are infectious diseases caused by invisible living creatures, or are they caused by something else? To know for sure, in scientific terms, we need to prove a relationship between a cause and an effect. And that is not easy. Noticing a few tendencies and making a few guesses is not enough. Without plenty of detailed exploration, careful experimentation, and thorough thinking, there will always be doubts and objections. In the case of germs, the process of proof took centuries. Most scientists — including most doctors and biologists — doubted the existence of living germs until the late 1800s. The famous nurse Florence Nightingale, for example, did not believe in them. What follows is a brief story of a long struggle to prove the existence of invisibly small, living agents of infectious disease. This is the story of the discovery of germs.

THE MICROSCOPE

In the year 1600, the world was filled with creatures great and small, and the smallest were the gnats and fleas, as they had been since history began. Every once in a while someone — perhaps a squinting nearsighted person, or someone with a weak magnifier — would observe tiny specks on a rind of cheese, or on someone suffering from "chronic itch," and he or she would claim that these almost-invisible motes were actually alive. But they offered no proof or even convincing evidence, the Authorities denied the existence of living specks, and everyone else ignored these wild claims.

In the year 1600, humanity was still suffering from sporadic storms of deadly epidemics, with no clue about the true cause, and little power to evade them. London had lost over 20,000 of its residents during an outbreak of bubonic plague in 1563, and the plague was endemic in Constantinople for most of the 1500s. Neither the ancient Greeks nor any subsequent scientists had made much progress in discovering the ultimate cause of such natural disasters, much less in devising any kind of cure or prevention. Mankind was still helpless against the Wrath of Pestilence.

But 1543 was the year Copernicus published his theory about the earth circling around the sun, and by the year 1600, the Scientific Revolution had begun.

The Two-Lens Trick

If you have a transparent sphere (perhaps a glass bowl filled with water, or a large glass marble), and you look through it at something on the other side, the object you are looking at appears enlarged. This effect has been known for thousands of years. "Reading stones" and magnifying glasses are just reduced versions of transparent spheres. These simple magnifiers can be fun and beautiful toys to play with. But they only make things look a few times larger at most, and for most of history, they were useful mainly as a reading aid or an assistance for people with aging eyesight.

By the year 1600, Dutch spectacle-makers had become proficient at grinding and forming a variety of quality glass lenses, both concave and convex. One can imagine children (or adults) wandering through the workshops, playing with the lenses, and trying to see what happens when you look through two lenses at once. Many of these people probably found only blurry colors, and quickly gave up. But a few of them discovered that if you arrange the lenses just so, you are treated to something remarkable. Try holding a concave lens (maybe someone's corrective eyeglasses) over your eye, and hold out a weak convex lens (i.e., a weak magnifying glass) in front of it. Try lining them up in your sight and looking through the pair of them at something far away. As you adjust how far away you hold each of them, you might discover that they make the distant object look closer and larger. Or look at a fly with a strong magnifying glass, then place a weak concave lens over your eye. With the right adjustments, you should be able to make the fly look even larger than you could with only the magnifying glass. Using a *pair* of lenses in the right way, you can extend the power of a simple magnifying glass. You can upgrade a toy into a powerful optical instrument. These discoveries were apparently made by several people independently within a decade or so of 1600. The first written documentation is a patent application filed by Hans Lippershey in 1608, but two-lens magnifiers may have been in use as early as 1590.

In about the year 1600, a scientific giant in Italy learned of the new Dutch two-lens magnifiers. This visionary man instantly saw the scientific potential of both kinds — the far-seer and the small-seer. He figured out how to manufacture his own optical instruments, and then he used them to see things that no human eyes had ever seen before. Galileo Galilei was most interested in the ability of the "far-seeing" device to extend his eyes into outer space. He had devoted his life to proving the Copernican theory of the solar system, and he realized how important this new instrument could be in supporting his case. But he also rejoiced in the tiny wonders revealed by his "occiolino," or "little eye." Putting a housefly in front of his "little eye," Galileo observed the fly's bizarre compound eye, and he discovered how the fly could walk on walls and ceilings without falling off. He later gave one of his "occiolinos" to Federico Cesi, a young friend. This youth admired Galileo's new world-observing, experiment-trying spirit, and at the age of 18, he gathered together some like-minded friends and founded his own club. It was probably the first "science club" in history. Cesi named his club the *Accademia dei Lincei*, using the sharp vision of the Lynx as a symbol for seeing nature clearly, and he devoted the rest of his life to his clear-seeing academy. Galileo later became a proud Lyncean, and it was members of this academy who gave Galileo's new vision-extending instruments their modern Latin names. To the "far-seer" the Lynceans gave the name "tele-scope," and to the "small-seer" they gave the name "micro-scope."

In the years following Galileo's death, the telescope became very popular and underwent rapid development. The microscope did not. The Lyncean Academy produced a few drawings of sights seen under the microscope, but sadly it evaporated shortly after the death of its founder. The microscope had not revealed new worlds, and it remained a novelty, a slightly-more-powerful magnifying glass, for decades.

The First Microscopists

Galileo had helped to plant the seeds of an intellectual revolution, but these seeds took a while to sprout. In Italy at least, Authority still had firm control over the "proper" views of how the world works. Galileo's imprisonment for contradicting Doctrine was still a fresh memory, and everyone else with a desire to explore the natural world and to see it clearly for themselves had to keep their mouths shut, or risk a similar fate.

But around the year 1660, a new crop of Galileo's intellectual descendants began to grow. Here and there in scattered corners of Europe, a few rogue explorers started adding lights and mountings to their microscopes, they started making careful, systematic explorations with their improved instruments, and they started discovering new truths. At least two of these pioneers figured out how to use a microscope to look inside a body and how to perform tiny dissections under a microscope, and they invented microscopic anatomy. At least two of them discovered the capillary blood vessels that link the ends of arteries with the ends of veins, thus providing the conclusion to a different story: the discovery of the circulatory system. Several of these microscope pioneers were talented artists, and they created beautiful color illustrations to show others what they saw. These five pioneering microscopists worked mostly independently, but sometimes they met or corresponded, and they were all active around the same time, from about 1660 to 1690 — two of them in the Netherlands, two in England, and one in Italy.

The Italian was Marcello Malpighi. He mapped the anatomy of the silkworm (previously thought to be too small and simple to have internal organs), and he helped to discover the lymphatic system (which was later discovered to play an important role in protecting us from germs). He is also the one usually given credit for discovering capillary blood vessels. The Englishmen were Nehemiah Grew and Robert Hooke. Grew used the microscope to search for fundamental similarities between plants and animals, hoping to discover unity in God's Creation. He found it in the realization that flowers are actually a plant's sexual organs. Robert Hooke was primarily a physicist, but his 1665 picture-book of sights seen through the microscope was arguably the first scientific bestseller in history. Perhaps one could say it went viral. Hooke's *Micrographia* included 38 plates and gave the astonished public its first view of a monster-sized gnat, a flea, and a louse, clutching a human hair. And those specks on cheese? They really are tiny bugs: cheese mites. (The specks on itchy patients are scabies mites. Those would be discovered a few years later.) Of the two Dutchmen, one was Jan Swammerdam, who performed tiny dissections and made beautiful paintings of them, and like Malpighi helped to invent microscopic anatomy and to discover the lymphatic system.

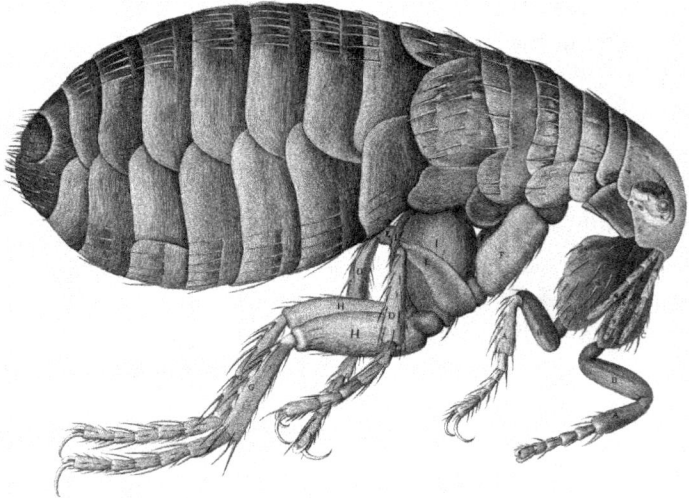

Robert Hooke's Illustration of a Flea

Robert Hooke's *Micrographia* was also the inaugural publication of a new science club in London: the *Royal Society*. Federico Cesi's devotion to seeing nature clearly had faced opposition from Authority, and this opposition may or may not have played a role in his Academy's downfall. The new science club in London defied Authority publicly in their motto: *Nullius in verba*, or: *Take nobody's word for it*. The Royal Society's membership included Hooke and most of the other pioneering microscopists, it would go on to include many of the subsequent doctors and scientists in this story, and the Royal Society remains a prestigious institution today.

Antony van Leeuwenhoek

The remaining Dutchman was Antony van Leeuwenhoek, and it was Leeuwenhoek who made the first major discovery on the way to understanding infectious diseases. (If you say "Lay-wen-huck," you might not sound like a native Dutchman, but you'll be pretty close.) Leeuwenhoek was a businessman by profession, but he spent all of his free time making and using microscopes as a personal hobby. Unlike the others, he did not use the two-lens trick. Leeuwenhoek chose a different design of microscope — a single-lens design that amounts to a very tiny but very powerful magnifying glass. You can see how it works using a small clear marble as a lens. Here's a "Leeuwenhoek microscope" I made by bending a paperclip into a handle for a marble:

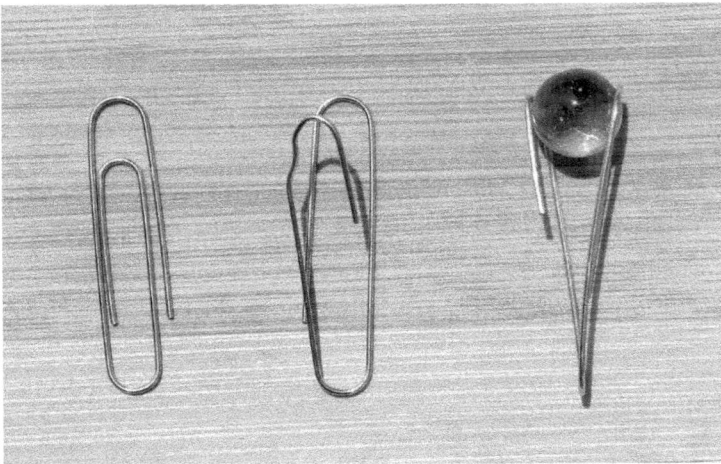

A Homemade Marble Magnifier

With such a device, you can make things look quite large by holding the marble up close to your eye, and holding the thing you want to look at close to the other side. You may need some strong light in front of you to obtain the best pictures. I took a picture of a bee's wing by taping the wing to a window on a sunny day, holding the marble up in front of it, and holding the lens of my iPhone camera in front of the marble. After a little juggling, I managed to capture this:

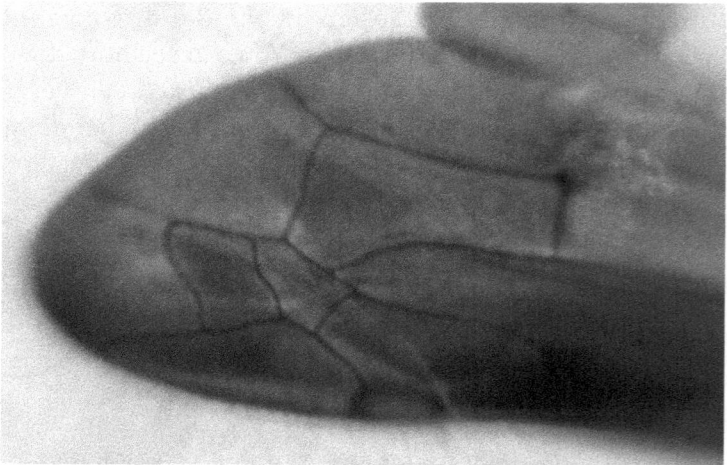

A Bee's Wing, Seen Through a Marble Magnifier

A marble magnifier gives you a better view of things you could see already, but it probably won't show you any new forms of life. The marble magnifies more than a large magnifying glass because the marble is so small. But to see really tiny things, you need really tiny lenses. Leeuwenhoek was able to see things far, far smaller than anyone had ever seen before because he figured out how to make really, really tiny lenses. Instead of the "two-lens trick," Leeuwenhoek used the "tiny-lens trick."

Leeuwenhoek perfected the art of grinding and mounting his tiny lenses, he invented ways to mount specimens against the lens, and he designed a way to make delicate movements of the specimen using fine adjustment screws. (Leeuwenhoek's movable mechanisms were the forerunners of the modern *microscope stage*.) Leeuwenhoek's hand-held microscopes were more awkward to use, but they were more powerful than the two-lens microscopes of his contemporaries, and with them he observed something remarkable, something completely new, something none of the others had ever seen or even dreamed of.

One of Leeuwenhoek's Microscopes

Photo courtesy of Douglas Anderson at Lens on Leeuwenhoek[1].

---

[1] https://lensonleeuwenhoek.net

We know about Leeuwenhoek's discoveries mainly through the numerous letters he wrote, mostly to the Royal Society of London. In one of these letters, dated 1674, he describes some beautiful, wispy, green streaks that he had observed one day in local lake water. From his careful description, we now know these wisps to be the lovely *Spirogyra* algae. But he also saw something strange around the wisps. Things were moving ... not just waving in the currents, but zooming across the window, or waddling strangely, or just whirling in circles. Some of them moved almost as if they were feeding on the algae. Swarming around the Spirogyra, Leeuwenhoek saw things far smaller than any speck of dust — but they swam and behaved just like living creatures! Leeuwenhoek had discovered minuscule animated animals, far smaller than any that had been known to exist before, swimming and dancing and zooming around the algae. (Leeuwenhoek wrote exclusively in his native Dutch, and the word he used for these things, the diminutive for "animal," is traditionally translated into English as "animalcule.") Leeuwenhoek wrote: "And the motion of most of these animalcules in the water was so swift, and so various upwards, downwards and round about that 'twas wonderful to see: and I judged that some of these little creatures were above a thousand times smaller than the smallest ones I have ever yet seen upon the rind of cheese."

If you own a microscope, it is not too difficult to find some of these creatures and see them for yourself. I found some critters in the algae growing in a horse's water barrel, and I took the following picture of them using only my iPhone and a hobby microscope. (If you don't own a microscope, it is not too hard to make one similar to Leeuwenhoek's. If you can handle a little light work with a glass rod and a flame, and most people can, I recommend the simple but elegant method explained by Patrick Keeling[2]. You can also buy a very inexpensive pre-made Leeuwenhoek-type microscope from Foldscope[3].)

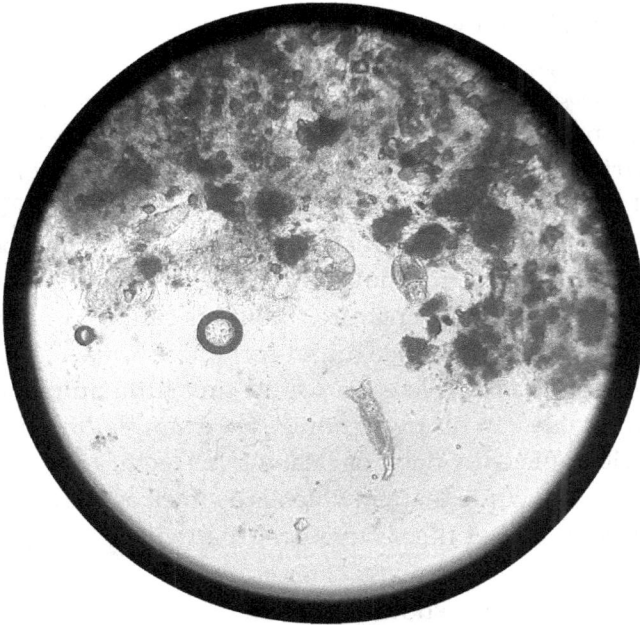

Rotifers Feeding on Algae

Galileo Galilei and Antony van Leeuwenhoek both made their own optical instruments and taught themselves the skill of using their new devices, and both of them discovered new worlds through their instruments. Thanks to the telescope that he built, Galileo was the first person in history to witness the phases of

---

[2]http://www3.botany.ubc.ca/keeling/microscope.html
[3]https://www.foldscope.com

Venus, the moons of Jupiter, and the mountains on the moon. Thanks to the microscopes that he built, Leeuwenhoek discovered a new world of his own, a world that nobody had known or even suspected before, a world filled with new, bizarre, alien creatures . . . right in our own back yard, but incredibly small. Leeuwenhoek discovered the microcosm.

Leeuwenhoek spent much of the rest of his life searching for, finding, and studying his little animals. Children delight in pets and in trips to the zoo; Leeuwenhoek delighted in the tiny zoo he had discovered inside his microscope. He searched every sample of water he could reach — lakes, the sea, his roof gutters, his well, the Delft canals, drainpipes — and he described new creature after new creature. From his careful descriptions, modern students of science might recognize *Rotifers, Vorticella, Euglena,* and *Hydra.* He even found tiny little animals *inside* other animals — in his own mouth, for example. In one paper, he described his daily habit of cleaning his teeth, and how it left behind "a little white matter, which is as thick as if 'twere batter . . . in the said matter there were very many little living animalcules, very prettily a-moving." He provided a few sketches, and described how they "shot through the water" or "spun round like a top." Leeuwenhoek was also the first person to describe the red oval corpuscles in blood and the wiggly little things in seminal fluid. To the best of my knowledge, however, he never associated any of these things with disease. Wherever he found them, "animalcules" were like the red blood cells: just natural residents in the place where they belonged.

Leeuwenhoek never stopped exploring his microcosm, and his diligence is one reason he holds a much loftier place in the history books than those in earlier centuries who claimed to have

20

seen living dust specks. Leeuwenhoek asked and then tried to answer as many follow-up questions as he could. How big are these little animals? They come in a range of sizes. Leeuwenhoek chose a few of them to analyze, and then he carefully compared them to known objects, and he calculated ratios. They were astonishingly small. "This creature is a thousand times smaller than the eye of a louse!" In rough terms, these new creatures were as much smaller than lice, as lice are smaller than human beings.

Where do the little animals come from? Leeuwenhoek tried an experiment. He collected some fresh rainwater straight from the sky in a thoroughly cleaned vessel, and found nothing. But when he checked again a few days later, little animals had appeared, along with a few bits of dust and lint. He doubted that they had magically appeared from nowhere, but he was a practical man, careful never to jump to conclusions. For now, the question of where animalcules come from didn't have a convincing answer.

What do the little animals eat? In a different experiment, Leeuwenhoek had prepared various "infusions" of ground pepper, ginger, cloves, nutmeg, and vinegar, hoping to discover what causes each of them to taste the way they do. Was the sharp taste of pepper due to sharp spikes or edges on the grains? After a little time, he discovered swarms of animalcules in these infusions ... and thus he accidentally discovered how to feed microbes. To this day, infusions of stuff from your lawn are a good way to grow crops of microbes to put under your hobby microscope. Or to feed to your pet fish. For many years, this entire new category of life was officially named *Infusoria*.

How does animalcule anatomy work? They have no muscles or limbs, nor any apparent brains or stomachs or hearts. Leeuwenhoek was fascinated by the organs of motion — or lack thereof. These bizarre creatures were clearly capable of rapid and agile motions, but they had no apparent means of propulsion or guidance. In a few of the larger creatures, Leeuwenhoek could see what looked like hair, or maybe tentacles sticking out from somewhere, but he never saw anything "normal" like fins or flippers.

Leeuwenhoek had discovered a new world of life, but without Robert Hooke, history might have written him off as a charlatan. Leeuwenhoek was perfectly willing to let people look through his instruments, to see what he saw. But nobody could see these incredible new animalcules except through Leeuwenhoek's personal, hand-made microscopes, and he refused to share his special techniques for making them. Some people questioned his honesty. Or his competence. Or at least his eyesight. But at the request of the Royal Society, and after three attempts, Robert Hooke finally managed to produce microscopes equal to Leeuwenhoek's. Hooke finally demonstrated to the Society that everything Leeuwenhoek had claimed was true, and he showed the rest of the world ways that they could see these things for themselves. Hooke brought proof to the world of the existence of microscopic animals.

The Zoo Under the Microscope

Birds. Reptiles. Insects. Mollusks. The inventory of creation needed a new branch now. Leeuwenhoek had added an entirely new wing to the world's zoological garden — the *Infusoria*.

Today, we call them more formally and more broadly the *microorganisms*. Or *microbes*, for short. The tiny occupants of this wing of the zoo come in all shapes and sizes, from the larger ones with a few visible parts to the tiniest simple living shapes (often straight sticks or round spheres). And they exist everywhere we look in the natural world, living and swimming in every kind of moist matter, even inside other living bodies.

"Normal" animals come in all shapes and sizes, each of them fitted to its own kind of lifestyle. Some of them make a living by feeding on other animals. If a larger animal kills and eats a smaller animal, we call them "predator" and "prey." Sometimes a smaller animal hitches a ride on or in a larger animal, and it makes a living feeding off of the larger animal without killing it. Then we call them "parasite" and "host." What about animalcules? Sometimes we find them living inside other bodies — could they be microscopic parasites? Could we explain disease by saying that some of these microbes find a way in and make a parasitic living inside of our bodies? Could parasitic microbes be the "seeds" or "germs" of disease? Perhaps. But even if we searched and found foreign microbes inside of us, how could they possibly be causing any harm? They are so small that they have no effect on our sense organs, so how could they cause any damage to our lungs or our stomachs? Some people could try to claim that microbes were the agents of disease, that the "something out there" that causes plagues and epidemics was actually microscopic life, but they were only guessing. There was no proof, nor even any evidence to support the claim. Surely microbes in a microscope are like the flakes in a snow globe — pretty to watch, but of no practical consequence?

For almost two centuries after Leeuwenhoek's discovery, this is how most of the world viewed microbes. They were known to exist, but they inhabited a different dimension. They were like mythical beasts, existing only in the world inside a scientist's microscope. It was a curious world, a fun-to-visit world, but also a useless and unimportant world, of no significance outside the window.

THE EXPERIMENTALISTS

How do microbes work? Do they eat? Breathe? Go to the bathroom? Can they get sick? What do microbes do for a living? Do they hunt? Graze? Do they make nests or burrows or hives? And where do microbes come from? How do they keep showing up in places that didn't have microbes before?

There was no easy way to answer these questions. For one thing, it was still difficult to see clearly what the microbes were doing. Leeuwenhoek's microscopes were hard to use and had very tiny "windows." Two-lens microscopes gave broad, clear images, but only at low power. If you tried to make a two-lens microscope too powerful, the images became washed out and blurry. So microscopists could only watch these creatures in a very limited way, like a wildlife photographer hiding in a tree and observing wildlife from afar.

But no matter how clear the view through the microscope had been, it wouldn't have been enough. We would never have learned all we know about normal animals by watching them from a distance. We learned most of what we know from millennia of hunting them, feeding them, repelling them, and domesticating them for pets and livestock, as well as studying them deliberately in laboratories. Now we needed to do the same thing with Leeuwenhoek's animalcules. To discover the habits and the lifestyles of the microbes, and to learn whether they have any significance for human beings, passive observation through the microscope was not enough. The next step in the journey towards understanding microbes was to find ways to cage them, to learn how to raise them, to poke them and prod them in a test tube, and to see what happens when you try different things. Learning how the microcosm worked required *experiments*.

During the same years that the first generation of microscopists was carefully describing and documenting the new world inside the microscope, another intellectual descendant of Galileo branched out in a different direction. Francesco Redi was another Italian, a contemporary of Malpighi, both of them living about a generation after Galileo. Galileo's ideas about putting the claims of Authority to the test had inspired hostility from Authority, but also courage and curiosity in a handful of the next generation. Redi was one of these. Redi specialized in medicine and biology, and he decided to start questioning established biological "facts" and putting them to the test. Francesco Redi was biology's first "Mythbuster."

Where Do Microbes Come From?

Redi started to build his reputation in the early 1660s by clearing up some confusions about poisonous snakes, their bites, and some alleged antidotes. Redi's contemporary Athanasius Kircher had been claiming that one could draw the harmful substances out of a snakebite victim simply by placing special stones against the victim's skin. These "snakestones" had been brought back from the Far East by missionaries, where they had allegedly been extracted from the heads of snakes, and they were supposed to work like magical poison magnets. Redi showed that they didn't work.

But a few years later, in Redi's magnum opus, he took a stand in an epic philosophical debate over the origins of life.

Where do living organisms come from? The origin of new people is obvious. Livestock, too. One could make a list of trees and flowers and vegetables that obviously grow from seeds or nuts that were produced by an earlier plant. In all of these cases, living things come from prior living things. And the earlier living thing is always of the same kind. Humans have babies. Rabbits give birth to rabbits. Oak trees grow from acorns produced by oak trees. Whenever a new life appears, there is always a "parent" of the same kind that created it. But maybe that's only true for large, sophisticated organisms. What about swarms of smaller, cruder vermin? What about maggots, grubs, slugs, worms, ants, wasps, and all other manner of squirmy, wriggly, creepy-crawly things? And what about algae? What about mold? Many of these things do not come from seeds or parents, at least not in a way that we can see. Many of them seem to condense spontaneously anywhere the conditions are right, out of any murky matter that can incubate them.

On one hand, it seems nice and tidy to just assume that *every-thing* is generated in a similar way — living things might be born, hatch from eggs, or sprout from seeds, but they always come ultimately from a previous organism of the same kind. On the other hand, any devoted observer of nature could make a list of cases to the contrary. Sometimes one finds little wormy things in a place that no wormy thing entered, such as maggots appearing from nowhere in a pile of dung. Sometimes a wasp will emerge from a lump or "gall" that grew from the side of a tree. Algae and slime grow in containers that once held nothing but pure water. Wherever the ingredients come together, living things are cooked up in nature's kitchen. Wherever life is possible, life is born. (At least one later philosopher, Georges Buffon, tried to theorize about a living "Force" that could flow through matter and assemble the elements of life into simple living things. Maybe that's where George Lucas got the idea.) But worst of all, the assumption that all life comes from previous life doesn't really solve the problem. To say that living things always come from parents begs the question: *Where did the parents come from*? It leaves you with an infinite regress.

As the decades and centuries went on, and the tide of Enlightenment science rose against the bulwark of religious authority, the question of the origin of life would also provide an additional battlefield for religious debate. The idea that living things always come from similar ancestors fits very nicely with the orderly, religious idea of a single moment of creation; the idea that life could be cooked up here, there, anywhere, seemed like the natural, messy, godless explanation.

So which is it? Can living things arise anywhere, naturally and spontaneously, wherever conditions are right? Or must every life form have parents, even swarms and slimes of the simplest and crudest kind? The only way to know for sure was through thorough investigation.

Athanasius Kircher came down on the side of spontaneous generation. He had somehow managed to obtain a microscope (possibly from the nearby Lyncean Academy, but nobody really knows) and he claimed to have found "worms" everywhere he looked. Based on these observations, he then told stories about an invisible "universal seed" that was spread throughout nature, and that sprouted into worms everywhere it found dung or diseased flesh or any kind of decaying matter. These worms could then give rise to higher and higher forms of life. (Kircher even gave recipes for creating things like bees, flies, and scorpions, but it's possible that he never even tried the recipes himself. Redi later showed that they didn't work any better than the snakestones.)

One of the places Kircher found "worms" was the blood of plague victims, and he wrote that "plague in general is a living thing." For this, I have seen him credited at least once as the first person to connect disease to microscopic "germs." But all Kircher had to offer were vague anecdotal reports of "worms," and speculation about "universal seed." It is possible that the "worms" he saw in the blood of plague victims were nothing more than tiny clumps of clotted blood, and his recommendation for avoiding the plague wasn't any more useful than his snakestones — he suggested wearing an amulet over your heart made from the flesh of a toad. (Oddly, however, I do wonder if one of his "worms" was really what we would now call a worm. He claimed to have seen worms in vinegar, and I have to wonder if he discovered, or re-discovered, the nematode worms we call vinegar eels. If he had documented his procedure more carefully and provided detailed descriptions of the worms, we might know.)

Francesco Redi came down on the other side of the debate. He believed that, in at least some cases, what seemed to be spontaneous generation really wasn't. And instead of simply asserting his belief, he tested it.

The First Controlled Experiment

Are worms generated spontaneously in putrefying meat? Redi decided to place some meat in a box where he could watch it carefully and control what happened to it, and simply observe what happened. Sure enough, worms appeared in the meat. But then they disappeared, and Redi wasn't sure where they had gone. So he tried again, and this time he closed the box after the worms appeared so they couldn't escape. The trapped worms searched for a place to settle down, and then ... they turned into eggs! Or at least something that looked like eggs: small, firm containers of some kind, like those made by silkworms, or by the "mealworms" you feed to your pet lizard. Redi wanted to observe the "eggs" separately, so he moved them to a glass jar and waited to see what came out, if anything. The eggs did in fact hatch ... into flies! By carefully controlling the circumstances of his observation, Redi discovered that the "worms" were actually a different form of fly. Flies can transform themselves from one shape to another, just like butterflies and silkworms. As we would now say, the "worm" or maggot is the *larva* of the fly, and it goes through *metamorphosis* to become the *adult fly*. Redi repeated the test with many different kinds of meat — more than seventeen different kinds by my count, including a variety of mammals, birds, and fish. (I have no idea how he obtained tiger meat.) Apparently, different kinds of flies prefer different kinds of meat, but in every case, Redi always observed the same sequence:

$$\text{worm} \rightarrow \text{egg} \rightarrow \text{fly}$$

Or as we would now say more formally:

$$\text{larva} \rightarrow \text{pupa} \rightarrow \text{adult insect}$$

Redi had discovered that certain worms transform into flies. But he still hadn't answered the original question: Where do the worms come from? Redi had noticed that flies were always hovering and buzzing around the meat — usually the same kind of fly that later hatched from the worms. Could the worms come from the flies somehow? Was there a full circle, a "life cycle," that began and ended with the fly? Redi wrote: "Having considered these things, I began to believe that all worms found in meat were derived directly from the droppings of flies, and not from the putrefaction of meat..." Less thorough thinkers may have been content with an assumption, but not Redi. As Redi said: "Belief would be vain without the confirmation of experiment."

But how could he create a truly convincing experiment, and not just an anecdotal observation? Suppose you place fresh meat in a container on your patio, and you cover it so maggots can't get in, and you watch what happens, and no maggots appear. Would that really prove that they can never appear spontaneously? Maybe they didn't appear for a different reason. Maybe you offered a kind of meat that maggots don't like. Maybe it was too hot, or too cold, or too bright, or too dark. Maybe the maggots need air, and they couldn't breathe in the sealed container. To really prove that maggots can't appear spontaneously in closed containers, you need to show that maggots *don't* appear, but you also need to show that they *would have* appeared, except for the fact that you didn't let flies in. To make a really convincing experiment, you need to have a *second* container, similar to your covered container in as many ways as possible, except that flies are allowed to enter one and not the other. The second jar shows that worms like the home you created for them, the first shows that they can't appear there if you close the door.

This was Francesco Redi's contribution to science. This was his own addition to the experimental tradition he inherited from Galileo. Redi designed the *experimental control*. He placed similar samples of meat in *two* containers, one open and the other covered with gauze, allowing air to pass through but not maggots. In the open container, he observed flies coming and going and the eventual appearance of maggots . . . which then turned into flies. In the covered container, he eventually observed maggots on the outside of the gauze, trying to get in, but no maggots or flies ever entered the jar, and no maggots ever appeared in the meat. The open container showed that the conditions were right for maggots, and the closed container showed that they can't appear inside if flies can't come in through an opening. With a controlled experiment, Redi had shown convincingly that worms are not generated spontaneously in rotting meat. They come from invisible eggs deposited there by flies.

A few of Redi's followers tried similar experiments with different bugs, and they found more insects that reproduce in the same way as flies. The microscopists also discovered some other insects that lay eggs that nobody had noticed before because the eggs were too small to see without the microscope. In his later work on parasites, Redi himself discovered numerous examples of invisibly small eggs. Kircher's belief in spontaneous generation had been based on anecdotal observations and vague theorizing. Redi's carefully controlled and documented experiments, with the support of similar experiments of others, placed him on much more solid ground when he expressed his new belief that "no animal of any kind is ever bred in dead flesh unless there be a previous egg deposit."

Redi's research also brought hope. If pests and parasites can appear spontaneously, then there isn't much that we can do about them. But Redi showed us that they can't just appear anywhere they want to. They follow rules, and if we understand the rules, then we can develop defenses against them.

Redi and his contemporaries had discovered many examples of invisible eggs and of the metamorphosis of worms into bugs. They had shown that in many cases where living things seem to appear from nothing, they actually still come from parents, it's just that there might be some invisible eggs or weird life forms in between. They had shown how insects can reproduce in a repeating cycle, like that of other animals, except maybe a little weirder and more complicated. But had they proven that *all* life must come from prior life? Even Redi wasn't convinced that it was true as a universal principle. He was unable to find an explanation for the appearance of gall flies and intestinal worms, for example, and he reluctantly allowed that these might somehow be spontaneously generated.

Incidentally, one of Redi's younger associates was a doctor by the name of Giovanni Cosimo Bonomo. Most microscopists up until now had either been professional scientists or microscope hobbyists. Bonomo was one of the first professional doctors to try to see if the microscope had any value in the practice of medicine. While studying "chronic itch," he applied his microscope to the skin and the sheddings of the patients, and he consistently discovered tiny creatures there. Bonomo provided sketches and descriptions of the creatures: They had many legs and "two little horns at the end of a snout." He also described the creature's eggs, and he explained the contagious nature of the itch as the passage of the creature from one person to another. Redi's friend Bonomo was the discoverer of the itch mite, now known as *Sarcoptes scabiei*. This discovery could be considered the first evidence for a link between a disease and a microscopic cause, more so than Kircher's "worms," but it was still hardly conclusive. Bonomo had a big row with the Church over it, and the idea that the mite causes the itch was pretty well rejected until the time of Pasteur. As with Leeuwenhoek's little animals, most people thought that the scabies mite was just an irrelevant tiny speck seeking its natural habitat.

Experimental Duelling

Now, what about microbes? Even if all "normal" life comes from eggs or seeds or parents, surely the same does not apply to the weird, tiny, rudimentary aliens in the microscope? Insects have heads and legs and wings, and maybe we can count them as complex "higher" organisms, but surely the tiniest, crudest forms of life are brewed up wherever there is a fertile infusion for them to swim in? We never see microbes lay eggs or give birth like normal animals. They simply appear or they don't.

So, are microscopic life forms cooked up spontaneously or not? Again, only careful observation and experimentation could tell us for sure. Leeuwenhoek had performed an experiment or two with his little animals, and he was reluctant to believe in spontaneous generation even for microbes, but he was an explorer in an uncharted realm, and like Redi, he was admirably cautious about jumping to conclusions. Not everyone was so cautious.

In 1748, an English Roman Catholic priest by the name of John Turberville Needham read about animalcules. He wanted to see if they could appear from nothing, so he took a trip to his kitchen. He cooked some gravy over a fire for a few minutes to kill any microbes that might be there, poured it into a bottle, corked it, and waited. After a few days, he opened the container, placed some of the contents under a microscope, and saw animalcules. He had offered nature a nurturing medium, and nature had brewed a fresh crop of microbes inside.

About two decades later, an Italian Roman Catholic priest by the name of Lazzaro Spallanzani caught wind of Needham's experiment, and he smelled something fishy. Spallanzani knew about Redi's work, and he admired Redi's carefulness and experimental technique. In contrast, Needham's work sounded sloppy. Had Needham cooked the gravy long enough? Had he adequately sealed the container? Microbes are pretty tiny — maybe they had found a way to sneak in through the cracks around the cork. Spallanzani conducted his own set of tests, this time with a variety of sealing methods and a variety of cooking times as comparisons and controls. Spallanzani picked up the tradition that Galileo had started and Redi had built upon, and he continued their pursuit of truth through experience and experiment.

In some tests, Spallanzani poured the gravy into a glass flask with a thin neck, and then he used a flame to melt the neck, fusing the glass itself over the opening and making a truly impermeable airtight container. These impenetrable flasks he boiled for an hour, and then he allowed them to stand for several days. When he broke them open and examined the contents, he found nothing but sterile gravy. Nature had not been able to form any microbes in those perfectly sealed containers. In another series of tests, he simply popped a cork into the mouth of the flask. He boiled these for an hour also, to ensure they were sterile to begin with, then waited. This time, the gravy teemed with life. Needham had apparently been sloppy and allowed microbes to sneak in around the cork. In further testing, Spallanzani sealed the flasks by fusing the glass, but then he only boiled the flask for a few minutes, as Needham had done. After examining these flasks, he also found microbes. Spallanzani had discovered that some microbes can survive brief boiling. Needham had performed a single casual test, and had come to the wrong conclusion. Spallanzani, using careful comparisons and controls, showed that if you *really* seal and sterilize a flask, then no microbes can appear inside until the seal is broken.

Needham did not take Spallanzani's rebuttal lying down. Needham was friends with Georges Buffon, the philosopher of the "Force," and together they tried to explain Spallanzani's results their way. For Needham and Buffon, life did not form truly "spontaneously," but under the influence of an invisible force that flows through matter and assembles the building blocks into simple living things. Spallanzani had boiled his flasks for an hour, instead of only a few minutes, and this must have somehow destroyed the "Force" in the gravy. In other words, the excessive boiling must have destroyed the ability of the gravy to generate life. They did at least take the time to repeat some of Spallanzani's experiments, and they also noticed that when they broke the necks of the flasks, there was a hiss of air as it rushed in to fill the vacuum in the flask. The flame used to fuse the glass had also heated and expanded the air inside, which left a partial vacuum inside the flask when it cooled back down again. For Buffon and Needham, this hissing was evidence that the air had been damaged or destroyed by the flame during the sealing process, and that the capacity for generating life had been damaged.

Spallanzani had much less to fear from Authority than his fellow Italians of a century earlier. He didn't need to worry about being punished for saying things that other people disagreed with. But that didn't mean that the road to clear understanding was now easy. More and more people, both careful people and sloppy people, were opening up with their opinions, and conflict and error were still dangers. Even the Royal Society was impressed with Buffon's theory of the origin of living things. But as long as observation and experiment, evidence and proof, were taken as the ultimate judge, then errors could be corrected, conflicts could be resolved, and truth could win.

Needham had reported the results of his initial casual experiment, and Spallanzani had given his more thorough rebuttal. Now Needham and Buffon had tipped the ball back into Spallanzani's court. Spallanzani spiked it. As before, he did not reply with speculation, but with thorough and careful experiments. Buffon and Needham had not bothered to wait and to see what happened to the broth after exposing it to fresh air. Spallanzani did. He left his broth to sit in his broken flasks, but now newly exposed to air. The broth that had formerly been sterile soon filled with microbes, showing that it had not lost the ability to support life after all. As long as the flasks remained unbroken, no microbes ever appeared inside, but whenever he broke the flasks and exposed the contents, they filled with microbes shortly thereafter. It was exposure to the outside world that made the difference, not damage to the broth. Spallanzani also invented a way to seal the flasks without creating a vacuum. With these flasks, there was no inrush of air when the necks were broken, yet they still remained sterile as long as the necks were unbroken. It was not "damage to the air" that prevented the microbes from growing in the sealed flasks. It was the seal.

Scientists and philosophers have always criticized and commented on the work of earlier scientists and philosophers. It can be very helpful to summarize and analyze what has been done by people before you, before you attempt to go further. And sometimes "duelling" with your contemporaries can also be useful. If I had been Francesco Redi or Lazzaro Spallanzani, I would probably have been annoyed and frustrated at having to debunk Athanasius Kircher or John Turberville Needham. And yet my own arguments might have grown stronger by it. As long as your opponents are at least partially devoted to finding truth, having to answer their objections can make your own case stronger. To this day, "peer review," in which scientists give feedback on each other's work, is an important component of the self-correcting mechanism of scientific discovery. It can be very effective when everyone involved is diligent, thorough, and respectful, and it can be invaluable for ironing out details. But it also runs the risk of becoming personal, contentious, and wasteful at times.

Where DO Microbes Come From?

If microbes cannot appear from nothing, then where *do* they come from? When you see two normal animals stuck together, they are probably mating. Sometimes you see two microbes stuck together. Are they mating? Spallanzani didn't immediately investigate this question himself. He was a busy man with diverse scientific interests. But his puzzle found its way to a fellow Renaissance Man in nearby Geneva. Horace Bénédict de Saussure attempted to corral some microbes in an infusion of hemp seed so that he could watch them closely, and he eventually came to the opposite conclusion. Each doublet was not two animals coming together, but one animal *splitting apart*! Microbes seem to have their own way of doing things in every area of life: eating, moving around, and reproducing. Perhaps their unique way of reproducing is not by laying eggs or giving birth, but simply by splitting into two halves.

Or maybe splitting is just a rare accident. Maybe, every once in a while, one of the speedy microbes comes zooming along without watching where it is going, and it smacks into another one, and breaks it into two pieces. That was the idea of John Ellis, who repeated de Saussure's experiments and observed that only a small percentage of microbes divide. He also insisted that the normal way for microbes to reproduce is by giving birth, because he claimed to have seen "children" microbes inside "parent" microbes, and even "grandchildren" microbes inside the "children" microbes.

Spallanzani heard about de Saussure's observation, and about Ellis's theory, and again something smelled fishy. He decided to return from his other pursuits and to focus his attention back onto microbes, and he showed once more what thorough, careful experimentation can do. He placed a droplet of infusion, bustling with microbes, onto a piece of glass, next to a droplet of pure water. With a delicate probe, he teased out a fine bridge of water connecting the two droplets, then carefully watched and waited. Eventually, a rambling microbe wandered across the bridge. Spallanzani quickly wiped away the thin bridge of water, trapping the microbe in a pond by itself. Spallanzani had invented a way to isolate a single microbe. He watched this loner carefully, and eventually he saw it split into two. Then he watched the pieces split again. He repeated the experiment and watched the division many times. For microbes, this is not a rare event caused by accident. *Reproduction by division* is routine procedure.

Where do new microbes come from? That question now had an observable answer, a demonstrable alternative to spontaneous generation. At least sometimes, they can multiply by splitting themselves. But there was still the question of where the parent microbes came from, and it had not yet been proven that microbes *cannot* self-assemble. Maybe they can generate in more than one way. After Spallanzani, the debate over spontaneous generation would flare up one final time, before being laid to rest, more or less permanently, by an intellectual great-grandchild of Galileo.

The Intellectual Tradition

I think you could argue that there is one living thing that is spontaneously generated, and that is an *idea*. Ideas and philosophies can grow and evolve from their predecessors like any living thing. But every *original* idea has to appear for the first time somewhere. It has to be born in the mind of a creative thinker, by an act of will. If there is some kind of "living force" capable of assembling prior elements into new and unique creations, that force is human concentration. Human thinking is what assembles observations and prior ideas into new ideas. The knowledge of humanity grows by addition, each new thought being created and integrated at some place in the world and some point in history, by the effort of one mind. I am not a historian, and I don't know how much credit to give to Renaissance thinkers prior to Galileo. But when recognizing giants of originality, Galileo must be in the first rank. In this one man was born a philosophy that you should look with your own eyes, put two and two together with your own mind, and test the assertions of Authority by your own experiences and experiments. And this intellectual bonfire ignited chains of sparks that can be traced through all subsequent intellectual history. In biology, Galileo ignited a tradition that passed to Francesco Redi, the Mythbuster who invented the controlled experiment, and then to Spallanzani, the man who cleaned up other people's messes and showed us where new microbes come from.

The illuminating torch of careful observation and experimentation passed from Galileo to Redi to Spallanzani, and then it passed across the Alps from Italy to France. It passed to a successor who gratefully hung Spallanzani's portrait on his wall, improved Spallanzani's experiments, and showed how microbes move the world. This man gave us the first *germ theory of disease*, thoroughly backed by evidence. He also delivered the final coup de grâce to the theory of spontaneous generation of microbes, and convinced the world that it is possible to defend ourselves against them. And then he gave us the weapons with which to do it. This man was Louis Pasteur.

Prelude to Pasteur

While Louis Pasteur was growing up, several more people made minor achievements of their own, and helped to lay the foundation for Pasteur. One of them provided a powerful upgrade to the primary tool of the microbe-hunter, while others finally began to find concrete clues about the "something out there," the external agent that was causing disease. So far in our story, we have focused almost entirely on the discovery of microbes, and we have said nothing about disease. Except for the itch mite, there has been no solid evidence for any connection between microscopic organisms and human diseases. But that was about to change.

The First Germ?

Who should receive credit for making the first connection between microbes and disease? Establishing priority in scientific issues can often be a tricky business, with plenty of room for nuance. In the discovery of germs, the credit for establishing the first *correlation* between a disease and a microscopic life form, should probably go to Redi's friend Giovanni Cosimo Bonomo. The credit for establishing the first solid example of a disease having a living external *cause* should probably go to one of Spallanzani's students, Agostino Bassi.

Bassi studied law to satisfy his parents, but he was fascinated by the natural world, and he took many classes in science, math, and medicine in school, some of them from Spallanzani. He began his adult life with an office job, but he eventually quit that job to take up a full-time career in agriculture. He applied his interest and skill in careful experimental science, and he wrote several authoritative and helpful books for growers. He helped shepherds to understand scientific ways to care for sheep, and he helped growers to produce quality potatoes, cheese, and wine. And for over 30 years of his adult life, he was perplexed by a bizarre disease of silkworms.

Sometimes the silkworms populating the mulberry orchards of silk-growers had epidemics of their own and died in droves. Sometimes their corpses became covered with a weird, white, somewhat powdery crust. Bassi tried to apply the methods of careful experimentation to the problems of the silkworm, but without much success at first. When he was almost ready to give up on the problem, he noticed something. Most doctors and scientists at the time believed that diseases developed internally, due to internal disturbances or imbalances of some kind, and Bassi had been proceeding on that assumption. But, perhaps inspired by his predecessor Redi and his carefully controlled jars, he then realized that an outside agent was necessary. Isolated worms did not get sick. Something had to enter their bodies from the outside.

Did it have something to do with the crusty coating? He tried pricking healthy silkworms with a little of the powdery coating taken from a dead silkworm. They quickly died. Often they would then develop a crusty coating of their own. Bassi found that the same thing happened if he tried pricking the pupae, or if he pricked adult moths. This powder was toxic to silkworms of any stage. But it was more than toxic. It could grow into new powder. It was reproducing. It was alive.

Bassi realized after examining the powder through a microscope that it was a kind of fungus. But this one was a *parasitic* fungus. Most common molds and mushrooms feed on things that are already dead, but this parasitic mold killed a living animal and then fed on the corpse. Bassi had discovered the first example of a contagious disease being spread by the passage of a microscopic parasite from one host to the next. He never managed to give a detailed description of the fungus or to identify an exact species. (Unlike Leeuwenhoek, Bassi did not have very good eyesight. His failing vision may have been one of the reasons he resigned from his office job.) But Bassi gave to history the first *parasitic theory of disease*, backed by evidence. Bassi had shown for the first time that an invisible life form (or at least its invisible seeds or spores) could cause disease in a larger organism. For this, Bassi has his own place in the honorable intellectual tradition. The species of fungus was later identified, and named *Botrytis bassiana* in Bassi's honor. And Pasteur would later hang Bassi's portrait on his wall as well, alongside that of Spallanzani.

Hospital Problems

In the 19th century, most women gave birth at home, where they were attended by midwives, and very few died. However, those without money, or with additional health problems, would often give birth in the charity ward of a hospital. The need to go to the hospital must have been terrifying, because women would frequently develop severe fevers while they were there, and many would die. Death rates over 10% were not uncommon in obstetrics wards. It was called "childbed fever" or "puerperal fever." It was awful, but nobody knew why it was happening or what to do about it, and so it continued. Outside the hospitals, almost no women ever got childbed fever.

44

In 1844, 26-year-old Hungarian-born Ignaz Semmelweis graduated from the University of Vienna as a physician. Two years later, he was appointed "First Assistant of Obstetrics" in Vienna General Hospital, and he noticed something curious about the death rates of women in his hospital.

Vienna General Hospital had two divisions, and one of them had a death rate that was about three times higher than the other. Other people had noticed this before, but nobody had been able to make much of it. Semmelweis was appalled by the deaths, and he decided to attack the problem. He began the laborious task of conducting autopsies of every woman who died in the hospital. His hospital had been keeping detailed statistics since it opened in 1784, so Semmelweis also started pouring through the records to see if he could find any interesting differences between the two divisions of obstetrics at his hospital.

There weren't any, except one. The conditions and practices were virtually identical between the two wards, except for the people. The ward with the higher death rate was being used for teaching purposes, and the women there were being attended to by medical students instead of midwives. Were the medical students somehow giving the women the deadly fever? Like his predecessors, Semmelweis couldn't understand it.

In 1847, as if Semmelweis didn't have enough death around him, his friend the forensic pathologist was accidentally cut during an autopsy, developed a fever, and died shortly thereafter. But death does not stop the pursuit of knowledge. Semmelweis reviewed the autopsy of his friend, and discovered similarities to childbed fever. That's when it clicked. Both medical students and forensic pathologists handled cadavers. It seemed as if something from the cadavers had been introduced into the pathologist's cut, and was being introduced into pregnant women by the medical students, and this something had made them all sick. Most doctors at the time attributed childbed fever, like most diseases, to some kind of internal flaw. Semmelweis realized that this hospital disease was being caused by an *outside agent*, something originating in cadavers and being transferred by contact. Semmelweis called this whatever-it-is "cadaverous particles."

In those days, medical examinations were done without gloves. Doctors and medical students would wash their hands in soap and water, as you might wash your hands before dinner, but that was all. And that was not enough to remove the "cadaverous particles." Semmelweis could tell that it wasn't enough, because the smell of rotting corpses remained on the medical student's hands.

Semmelweis set out to find a way to completely cleanse the hands of anyone who examined patients. He experimented with various cleansing agents, and settled on a solution of chlorinated lime (similar to common chlorine bleach) as most effective at removing the cadaver smell. He obtained permission from his boss to change the rules for the medical students, and instructed them to wash their hands in this solution between autopsies and medical examinations. The death rates dropped dramatically. Semmelweis had saved lives.

But then the rates spiked again a couple of times, and Semmelweis realized that the contamination can come not only from cadavers, but also sometimes from the patients themselves, if they had gangrene or some other smelly, pus-producing "rotting disease." Semmelweis concluded that the cause of childbed fever was not strictly cadaverous material, but any kind of "decaying animal organic matter." When any kind of animal flesh rots, that rot can cause rotting disease if it enters a healthy person. Semmelweis now instructed students to wash their hands in bleach between every patient and not just when entering the ward, and to isolate any patients that risked contaminating the general environment. Death rates dropped back down to low levels, and stayed there.

Semmelweis never found recognition in his lifetime. Caring for patients was his primary job, not scientific research, and he never identified or described any kind of germ. He could only call them "particles." He never gave any evidence about the nature of these particles or how they caused disease. In scientific terms, he only established a correlation between hand-washing in disinfectant and lower death rates. Most doctors still believed that childbed fever was due to internal causes, and a few were even offended at the idea that their hands were unclean. But the statistics were clear. Everywhere Semmelweis went, death rates due to childbed fever plummeted.

In the 19th century, Western Civilization no longer had one dominating central Authority telling everyone what to believe, and punishing those who disagreed. But the specter of Authority could still cast its shadow over clear thinking, even among scientists. Sometimes it is easier to insult an outsider or to dismiss him as a crank than it is to defend your own ideas. Sometimes it is easier to wrap yourself in tradition than it is to formulate a rational argument. It is too easy to lose the thrill of exploration, of seeing the world clearly, and of thinking for yourself that energized Galileo's descendants. It is too easy to settle into a desiccated bureaucratic mindset. (I fear there are dark streaks of Authority running through colleges and medical schools to this day.) Galileo had had to fight against Papal Authority. Redi and Spallanzani had had to fight against sloppiness. Semmelweis had to fight against Authority in his own profession.

Tragically, Semmelweis lost. Ignaz Semmelweis eventually went mad, was forcibly confined to a mental institution, was beaten by the guards, and died from an infection shortly thereafter.

The Next Generation of Microscopes

In the 18th century, the world had more or less given up on Leeuwenhoek's single-lens design of microscope because it was inconvenient to use and it was hard on the eyes. For the best results, one had to have sharp natural eyesight (which Leeuwenhoek probably had). On the other hand, the two-lens design started to show colored fringes above a magnification of 30 or 40 or so. In effect, when you demanded too much strength from the lenses, they started to act like prisms and turned the images into wet watercolor paintings. The two-lens microscopes were useless at high power.

In 1824, an English wine-seller and natural history enthusiast named J.J. Lister decided to see if he could make each lens out of two different kinds of glass, the second one correcting the mistakes of the first. He succeeded, and thus invented the first *achromatic lens*. These do not act like prisms, even at high magnifications. With a pair of achromatic lenses, two-lens microscopes could give comfortable, roomy viewing windows with crystal-clear images, at magnifications well beyond even Leeuwenhoek's microscopes. Thanks to Lister, the next generation of microscopes was born, and the optical microscope took on the modern design that we know today. These powerful modern microscopes could now make out structures (were they internal organs?) inside the larger animalcules, and they could give a much better look at the tiniest of the animalcules, such as the ones Leeuwenhoek had found in his teeth. In the ocean of whales and minnows, we could finally get a good look at the minnows.

By 1850, with the aid of the improved microscopes, the short bestiary of microbes had grown into a large catalog, organized crudely into categories by size and shape. In Pasteur's day, grouping and naming microbe families was still a messy business, but there were a few broad categories. There were the huge "whales" of the microbe world with complex shapes and visible internal organs — these were the *protozoans*. There were also swarms or "colonies" of tiny rods (*bacilli*), tiny spheres (*micrococcus*), and tiny spirals (*spirillum* and *spirochaeta*). Modern readers may be familiar with the names *streptococcus* and *staphylococcus* — those are tiny spheres that grow in long strings resembling chains of beads, and tiny spheres that grow in clusters resembling clumps of grapes. Collectively, all of these tiny "minnows" with simple shapes came to be known as *bacteria*.

In 1850, armed with a heritage of careful experimentation and a new generation of microscopes, Louis Pasteur was about to show for the first time what these microbes could do.

Pasteur and the Lives of Microbes

In the early 1850s, Louis Pasteur finished his education in chemistry and physics, and he began to climb the ranks of French Academia. In 1856, French sugar-beet growers were having problems with their ferments, and they decided to seek advice from science. (Apparently, sugar beets are good for making rum.) They enlisted young Pasteur to attack their problem. "Fermentation" was the name given to the foamy process that turns sugar into alcohol, and who better than a chemist to study what happens when fermentation goes wrong?

Wine and Milk

To ferment a batch of alcohol, the beet growers would honor their traditions, boil their beets, follow their recipes, and mix up their mashes in the prescribed way in the prescribed containers. And they would usually be rewarded with a batch of fermented beetroot alcohol. But sometimes the ferment would go bad, and for no apparent reason the distillers would end up with a batch of sour juice instead of the desired alcohol.

Pasteur, trained as a chemist, easily discovered what was making the bad batches taste sour. They contained the acid of sour milk. Or as chemists say now, *lactic acid*. But where had it come from? Somehow there had been a different conversion process — one that produced lactic acid when it should have produced alcohol.

What was the cause of the different outcomes? Pasteur decided to see if he could find any interesting differences under the microscope. From a successful ferment, he scooped up a little of the froth, and placed it under the lens of a microscope. When the picture came into focus, he saw, not animalcules, but something new. There were weird globules, just sitting there. In the microbic ocean of whales and minnows, these oval globs were quite small, but still a few times larger than the minnows. Pasteur had discovered a new globular kind of microorganism a little larger than bacteria. He had discovered the *yeasts*.

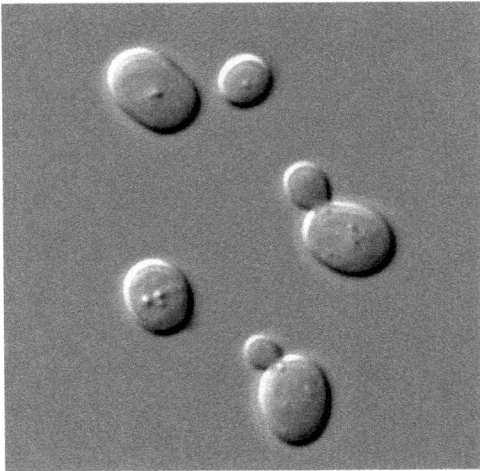

Budding Yeast Cells
(Public domain photo from Wikimedia Commons)

Actually, he had re-discovered the yeasts. In 1835, the French physicist Charles Cagniard de la Tour had examined brewer's yeast under a microscope and observed the globules. He had even realized that the globules were alive, because he saw them reproduce, but they reproduced in a strange new way. When a microbe reproduces by division, it pinches itself across the middle to form two equal halves. Cagniard de la Tour watched tiny little nubs start on one side the yeasts, grow larger and larger, and eventually detach themselves. Cagniard de la Tour had discovered that yeasts reproduce by *budding*. And this reproduction of the living yeast probably had something to do with the process of fermentation ... but what?

In his good ferments Pasteur rediscovered the yeasts, and the budding. What would he find in the bad ferments? From a sour ferment, he scooped up a little of the froth and placed it under the lens. When the picture came into focus, he saw tiny rod-shaped things, even smaller than the yeasts. Were they a new kind of living bacilli?

Pasteur realized that he needed to try to grow them. If they were living bacteria, he should be able to find a healthy soup that they could live in, and then he should be able to raise a crop of them in the laboratory. But he also needed a *clear* soup, so he could watch them while they grew and lived. He tried pure sugar-water, but the rods didn't like that. Eventually he came up with a recipe for a clear broth that made the strange rods grow. Whenever he introduced the rods into the broth, they multiplied, and the more they multiplied, the more acid of sour milk appeared. Not only were the rods alive, they were *manufacturing the lactic acid as they grew.*

Why did they make acid? Did it serve a defensive purpose, or was it simply a waste product, the microbe equivalent of urine? In any case, the lactic acid was coming from the bacilli as a natural part of their living processes, just as the alcohol was coming from the yeasts as a natural part of their living processes.

Wherever there were yeasts, there was alcohol, growing in quantity as the yeasts grew and reproduced. And wherever there were these bacilli, there was the acid of sour milk, growing in quantity as the bacilli grew and reproduced. A culture of yeast, or of bacilli, is a living chemical factory.

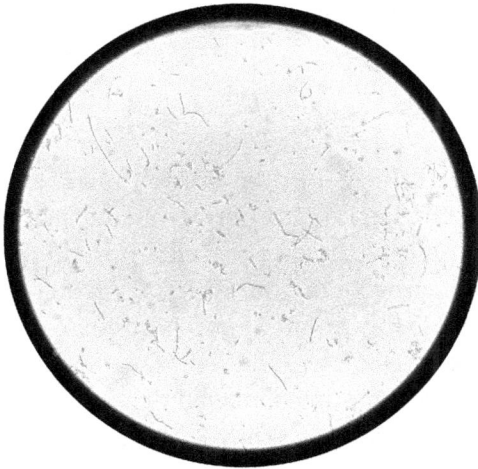

Bacilli in sauerkraut from the author's refrigerator

By isolating the bacilli, figuring out how to cultivate them, and studying them separately in their own test tube, Pasteur the experimentalist had finally shown that microbes are meaningful. One microbe alone might be as irrelevant as a speck of dust, but en masse, microbes can change the world. Swarms of them can manufacture so much liquid that it makes a difference to us, in our familiar world outside the microscope. Brewers have been unknowingly relying on microbes (or have been unknowingly frustrated by them) all along. Converting sugar into alcohol is what yeasts do. And converting it into lactic acid is what these bacilli do. Now we know Pasteur's rods by the name *lactobacilli*, and *lactic acid fermentation* is their lifestyle. Along with other *lactic acid bacteria*, they are an important resident in the digestive tracts of many animals, and they are used to ferment foods from yogurt to kimchi. Today, we generally think of them as "good bacteria," and they are included as a major component of many popular probiotic products.

The Spoilage of Food . . . and of People

But there were still exceptions. Sometimes Pasteur would attempt to grow the bacilli and perform lactic acid fermentation on purpose. (You can do this yourself: chop up a cabbage, pour some salt over it, crush it in your hands until it gets mushy, place it in a jar and press the air out of it as best you can, and then let it sit in the dark. A week later, you'll have sauerkraut, full of lactic acid bacteria.) But sometimes instead of sour acid, there was the smell of rancid butter. Was this yet another kind of "fermentation"? When Pasteur examined this new smelly "ferment" under the microscope, he found a new kind of microbe. Three different chemical products, three different microbes, three kinds of "fermentation."

And then the global power of microbes must have started to come together in Pasteur's mind. Many "chemical" processes — the fermentation of foods and beverages, the souring of milk, and the rancidification of butter — are actually all the work of living microbes. What about the rapid spoilage of any unpreserved food? What about the composting of kitchen waste into soil? Where do all the leaves of fall go before the winter is over?

What causes corpses to disappear and leave only the skeleton? What if microbes are responsible for all of it? What if microbes are responsible for every kind of decay and decomposition that goes on in every kind of organic material everywhere in the world?

The souring of milk is a kind of fermentation, similar to alcohol fermentation, and also similar to *putrefaction*, the smelly decay of organic animal matter. And living microscopic organisms are responsible for all three conversions. Maybe all fermentation, all putrefaction, and all decomposition of all organic matter everywhere is the work of different kinds of microbes going about their normal lives. If so, then microbes don't alter the world just a little bit, they *terraform the globe*. In all the world, in all the soil, in the corpses of every plant and animal that nature has ever produced, some chemical agent must be responsible for "digesting" that matter, for causing it to decay, for reducing it to raw ingredients, and for creating the soil from which the next generation of life can emerge. If this did not happen, new life would not be possible. The globe would be a landfill packed with every dead plant and animal that ever existed. Something recycles the matter from every generation of life and prepares the nutrients for the next generation. That something is microbes, and Pasteur's vision saw it for the first time.

And what then about diseases? If microbes were responsible for the spoiling of food, could they spoil people as well? If microbes could transform the soil of the world, could they transform our insides, too? Maybe microbes weren't so innocent and blameless of disease after all.

In Pasteur's time, there were several "hospital diseases" that were becoming more and more problematic. Childbed fever was one example, but there were other smelly, colorful, pus-producing and fever-inducing hospital problems, such as gangrene, erysipelas, and pyaemia. As hospitals became more and more common in the 18th and 19th centuries, more and more of their patients had various parts seem to rot away, even while the patient was still alive. And more and more people began to see that these "rotting diseases" bore a certain resemblance to putrefaction.

If microbes are responsible for decomposing animal matter in the wild, could they also be responsible for causing living flesh to decompose? Could Semmelweis's "corpse particles" actually be living "microbes of decomposition," carried from corpses or from the pus of living patients to other patients? People in hospitals often have their insides exposed, whether due to traumatic injury, or deliberate surgery. Maybe these openings allow "microbes of decomposition" inside the bodies of hospital patients, and then the microbes try to decompose their insides? Maybe all of those fever-inducing and pus-producing hospital diseases were due to an input of microbes, or as we would now say, *bacterial infections*? ("Infection" derives from Latin for "to put in".)

Pasteur began to see the global power of microbes, both for good and for ill.

But where were all these microbes coming from? When there is a fresh wound, or an open container of sterile beetroot mash, the "microbes of putrefaction" or the "microbes of fermentation" appear without being added. No other substance touches the wound or the mash. The microbes just appear. How could they get there? This brings us to the final eruption of the spontaneous generation debate, in which Louis Pasteur finally settled the issue of where microbes come from.

The Swan-Neck Flasks

Pasteur, like Spallanzani before him, did not believe that microbes could appear without parents. But if that was true, then there was only one other possible explanation for the appearance of microbes in places where there were no microbes before. They must be able to fly. Well, maybe not fly. But maybe they can drift through the air like motes of dust? Most microbes are swimmers — can they stand drying out and float alone through the air? If not, could they hitch a ride in a droplet of mist or on a wet particle of dust?

Pasteur wanted to show once and for all that microbes cannot come from nothing. They must enter new areas by traveling through the air. But now Pasteur had a problem similar to Francesco Redi's. If he simply sealed a bottle and sterilized it, and no microbes showed up, would that really prove that they can't be generated spontaneously? His opponents could argue that the sterilization damaged the air somehow, or that the microbes need fresh air to breathe. Pasteur needed to figure out a way to allow fresh air to enter the bottle, but to keep dust out.

Pasteur's solution? The famous *swan-neck flask*. The broth would rest in a glass globe equipped with a long, narrow neck. The fine neck would allow air to pass through, but it would prevent wind from blowing dust or debris into the interior. To prevent dust from falling down the neck under gravity, the neck arced up and over and back down again. Dust would have to fall up to reach the interior of the flask, and dust can't do that. So fresh air could pass in and out through the long, narrow, twisty neck, but no dust could blow or fall in.

One of Pasteur's Swan-Necked Flasks

Photo from the Wellcome Collection, London

Pasteur boiled his broth in the fancy flask, and then he waited. After several days, despite the fact that the rich nourishing broth had been exposed to open air, there were no microbes. Could he be sure that the broth was actually healthy for microbes? In the spirit of Francesco Redi's controlled experiments, Pasteur carefully broke off the curved portion of the neck, allowing dust to settle in through the newly exposed opening in the top of the flask. A few days later, the broth was full of microbes. As long as the neck was intact, the broth would stay sterile, but as soon as he broke the neck and let dust fall in, microbes began to grow. The broth was a good home for microbes, but they couldn't appear there unless they could float in with the dust. The same thing happened when Pasteur tilted the flask — as long as the flask remained upright, no microbes appeared inside, but as soon as he sloshed some broth into the U-shaped part of the neck where some dust had collected, washing the dust into the broth, microbes began to appear in the broth. With controlled experiments, Pasteur had finally shown how new microbes can appear in formerly sterile places. They can float in with the dust in the air.

Pasteur even developed his own method for testing the air in different places to see how many and what kinds of microbes it contained. It was a sort of low-tech air purity tester. He prepared many sealed sterile flasks with broth. He sealed them by fusing the glass, as Spallanzani had done, and he did it while the broth was at full boil, so that there would be a vacuum inside when it cooled down again. These flasks were his air testers. He would take them into whichever air he wished to test, then break the necks. Air would rush in to fill the vacuum, and then Pasteur would simply re-seal them and incubate them. If the air had been full of microbes, the broth would soon turn cloudy. If the air had been clean, the broth would remain clear. He used such flasks to show that the air outdoors was full of microbes, but the air in a clean and calm storage vault was relatively microbe-free. He also climbed a mountain, stopping several times along the way, and showed that the higher he climbed on the mountain, the fewer microbes there were in the air.

Pasteur showed that microbes can travel through the air with the dust, and he finally convinced the world that microbes are not spontaneously generated. And like Redi's work with parasites, this gave hope to humanity. Microbes follow rules, and this means that we can try to control them.

Whenever you leave food lying around in the open, it quickly spoils. Yet you can keep cans of tuna and cans of soup on your shelf for months or even years. Why? Because microbes are (mostly) the cause of the spoilage, but they can't pass through the walls of the cans, and they also can't appear spontaneously inside. If microbes could appear spontaneously, we would have no defense against them. But Pasteur showed us that they can't, which meant that we could invent ways to sterilize and seal foods and beverages in airtight containers, and then fill our shelves with long-lasting preserved food. (Pasteur himself, in his earlier work with fermentation, had invented a way to kill yeast in wine without damaging the wine. He gently heated it, using a process we now call *pasteurization*.)

Pasteur had shown that microbes cannot appear in any place that they cannot enter. And this meant that we could defend ourselves against them. We could throw up walls and keep them out of places we didn't want them. And if we could preserve food by keeping microbes out of food containers, could we also preserve people by keeping microbes out of their interiors?

Antiseptic Surgery

J.J. Lister, the man who had invented the achromatic lens and revolutionized the microscope, had a son who revolutionized surgery.

60

As a child, little Joseph Lister absorbed his father's interest in natural science and in microscopes. Then he grew up and became a professor of surgery at the University of Glasgow. Like many doctors of the time, he lamented the state of his own profession. "Hospital diseases" were still widespread, and half of all surgeries ended in death. Then Lister read about Pasteur's work with microbes. Were hospital diseases being caused by "microbes of decay" entering open wounds? If so, then preventing them should be a simple matter of killing any microbes that were already in the wound, and then keeping the air and the surgical instruments sterile. Once the germs were gone, they could not spontaneously reappear in the wound, and if you were careful, you could keep new germs from floating in on the dust. Pasteur had given Lister a way to save his patients. But there was still a pretty big problem. Killing microbes is easy. But how in the world do you kill the microbes in a wound *without killing the patient, too?* Lister had to find a cleansing agent that could kill germs, but without damaging the patient's own flesh too much. The "diseases of decay" were sometimes called "septic" diseases, from a Greek word meaning "to rot" or "to make rotten." What Lister needed to do was to find a suitable anti-rotting agent, or *antiseptic*, one that would prevent the infection and the rotting, without harming the patient.

The processing of coal and the burning of organic matter produce a variety of tars and other related substances, many of which have preservative properties. One of them is creosote, which was named after smoked meat, and has long been used to treat railroad crossties and telephone poles to keep them from rotting. A closely related preservative, recently discovered in Lister's time, is a substance now called phenol. In 1865, Lister settled on phenol as an effective antiseptic agent. He developed a careful procedure for swabbing wounds with this disinfectant, and he started disinfecting his surgical patients while keeping his surgical room as clean as possible. His ideas and techniques were mocked at first, but in a few years, his surgical death rate dropped from 45% to 15%. A decade later, the rate was down to 5%. Joseph Lister saved countless lives, saved limbs from amputation, saved the King from appendicitis, and vindicated Ignaz Semmelweis. He was honored with titles, medals, and statues from several nations, and his name is now immortalized in medicine cabinets everywhere on the label of a popular mouthwash.

If you wander through a sparkling, sanitary, modern hospital, and you smell that "hospital smell," you are probably smelling phenol. That's the smell of salvation from rotting death, and you can thank Joseph Lister for it.

French Silkworms

"Hospital diseases" or "septic diseases" were now solidly linked to bacteria. Semmelweis's "cadaverous particles" were real, and they were alive. They were simply the "microbes of decay" that are found everywhere in decomposing organic matter. Modern readers may be familiar with "strep throat" or a "staph infection." Pasteur observed and described both streptococcus and staphylococcus bacteria, and he associated them in a general way with diseases of decay. They are spread throughout nature, and they cause disease when they accidentally get inside of a living body, where they attempt to decompose the living flesh while it is still in use. They can be deadly if allowed to remain inside. But under most circumstances, our bodies are quite capable of making sure they stay outside where they belong. And even if they get in, in the modern world we can usually eliminate them easily with antibiotics. Fortunately, these microbes are not "out to get us." Their way of life is to decompose organic matter, not to attack us specifically. When we have an infection, they are doing the right thing, just in the wrong place.

But what about more hostile microbes? Are there *parasitic microbes*, whose way of life *is* to feed on us? Are different kinds of plague or contagion due to different species of parasitic germ, spreading from host to host? Pasteur thought so, but he knew that he needed proof if he was going to convince the rest of the world.

While Lister was working out his new procedure for antiseptic surgery, the French silk industry was suffering from problems similar to those of the Italians a few decades earlier. French silkworms were being ruined by the millions by disease, and the silk industry's revenue had plummeted to a tiny fraction of what it once was. In 1865, the French Minister of Agriculture asked Pasteur to turn his attentions to the problem. Pasteur initially resisted. He was a chemist, with no medical training. However, he knew that if he was ever going to prove his germ theory, he needed experimental evidence, and silkworm disease might make a suitable test case. He eventually accepted the job. He set out to find the culprit, and pin the blame on the guilty germ.

Unfortunately, the diseases of French silkworms were a little like the Gordian Knot. Silkworms suffer from multiple diseases with different causes, and at least two of them were active in Pasteur's time. Sometimes the diseases affected the eggs, the worms, and the adult moths differently. Pasteur never did untangle the knot and conclusively link a single specific disease with a single parasitic microbe. But like Alexander, Pasteur found a way to cut through the knot and move on. With his microscope and his experiments, Pasteur showed that the diseases were transmissible and were due to living agents spreading from worm to worm. Even if he didn't know for sure which living agents caused which disease, he at least figured out a way to identify which creatures were sick and keep them from infecting the healthy ones. The following year, the growers enjoyed a splendid yield of silk, and life in the French silk industry could begin to return to normal.

The Germ Theory

The pioneers of the microscope had opened up the microcosm, with diligence and detailed documentation. With thorough testing, the experimentalists had shown that microbes are more or less "normal" living things. They might eat and reproduce in weird ways, but they still needed to eat and to reproduce.

But before Pasteur, microbes were still curiosities that existed only inside a scientist's microscope, as insignificant to humanity as specks of dust. Outside the laboratory, they were still invisible and unrecognized. With extensive carefully controlled experiments, Louis Pasteur had shown that microbes change the world. They consume and emit materials just like any other living thing, and their collective chemical processing is constantly reforming the globe. Microbes are not like flakes in a snow globe. They are like flakes of snow — each one may be tiny and insignificant, but collectively they can transform nature. Pasteur also showed us that different microbes do different things. They are not random chaotic clowns of nature. Each kind of microbe has its own anatomy, its own habits, its own needs, and its own effects, just like any other living species. And Pasteur finally convinced the world (or the vast majority of it, at least) that microbes cannot be spontaneously generated. Microbes must come from parents like any other living thing.

But do microbes actually cause human epidemics? Are the outbreaks that plague humanity due to microscopic parasites, passing from host to host? Microbes cause fermentation and putrefaction, including the "diseases of putrefaction" that appear in unsanitary hospitals. And some kinds of microbes seem to be responsible for the mess of illnesses that appear in silkworms. It was not a huge leap to assume that infectious human diseases are caused by parasitic microorganisms, each disease by its own "seed" or "germ." Pasteur became convinced, and seeing the cause of infectious disease, he also saw the solution. He had shown that even microbe parasites cannot appear spontaneously inside a host. And if that was true, then humanity finally had a way to attack its deadliest problem. Humanity would finally have a defense against the Wrath of Pestilence.

Pasteur began promoting his "germ theory" and became a crusader against germs. He delivered passionate lectures to his students on the subject, and announced to the public: "It is in the power of man to make parasitic maladies disappear from the face of the globe, if the doctrine of spontaneous generation is wrong, as I am sure it is."

But it was still an assumption. There still had been no conclusive proof and no definitive link between human epidemics and microscopic parasites. There was still no knowledge of specific germs and no way to diagnose specific diseases, much less cure them.

By some accounts, Pasteur could be prone to streaks of arrogance and pettiness. I do not know. I cannot claim to be an expert on the life or personality of Pasteur. But most of what I have read about Pasteur written by those who knew him portrays him as a caring, thoughtful man, and a passionate defender of the truth when necessary. In any case, Pasteur's next project was not to immediately seek cures for humanity's illnesses but to defend the honor of France. When the Franco-Prussian War broke out, when the Museum of Natural History was bombarded, when Pasteur's new laboratory was threatened, and when Pasteur had to locate his own son in the mountains among the wreckage of his son's army unit, his sentiments turned against all things Prussian. So Pasteur turned his attention to what may have been a revenge project: making French beer better than German beer.

His national pride may also have helped to eventually turn Pasteur back to the problem of human diseases. After seeing the success of Robert Koch, a disease-hunting German, Pasteur would later rededicate himself to the hunting of germs and the extermination of disease. He would eventually provide mankind with weapons of mass destruction against several deadly diseases, and by showing how such vaccines could be made, he would inspire the creation of many more. In the saving of lives from deadly diseases, Pasteur would eventually become a titan ... but only after someone else first proved the cause of infectious disease. So for the time being, our story crosses the Rhine and passes from France into Germany. We turn from Pasteur to Koch, the careful, patient, clarity-producing, self-made scientist who linked specific epidemic diseases with specific microbial causes for the first time, and made germ-hunting into a science.

CAUSE AND EFFECT

I think the fictional detective Hercule Poirot would have admired Robert Koch. With rigorous "order and method," Koch was the man who finally proved beyond a reasonable doubt the guilt of the culprits behind three crimes against humanity. Koch was the man who finally convicted the first three germs of disease.

Koch, his Microscope, and Splenic Fever

In the late 1860s, a young Robert Koch graduated from a German medical school, repaired some soldiers during the Franco-Prussian war, married, and then settled down to a quiet life as a respectable country doctor. In 1871, on his 28th birthday, Frau Koch bought her husband a microscope to play with, and inadvertently launched his career as a germ detective.

While Koch played and practiced with his new microscope, the sheep of Europe were dying by the flock from a mysterious ailment that would turn their blood and their spleens black. Their spleens would swell with this blackened blood until they almost seemed to fill the torso. Occasionally shepherds would grow feverish, too, and break out in black sores. Sometimes they would die. Certain fields and pastures of Europe had become deadly ground — any flock of sheep or herd of cattle that grazed upon these cursed lands would become feverish and die with blackened blood and swollen spleens. These fields could kill, even if they had been untouched for years. Was there something different in the soil of these cursed fields? Something wrong with the local water supply? The local weather? This mysterious affliction went by a variety of names: splenic fever, the black disease, or in France *le charbon*, the charcoal disease. Now we call it *anthrax*.

As he attended to his patients with one disease or another, Doctor Koch would sometimes despair over his ignorance and his inability to do anything but dispense empty words and palliative pills. Even in Koch's day and age, doctors still didn't understand what really caused all that interference with people's breathing, or their digestion, or their blood flow. People got sick, and even doctors didn't really know why. And without understanding, there was little they could do. It was very disheartening. But in his newfound hobby as a microscopist, Koch was growing in skill and technique, and he would put whatever he could find under the lens. So when anthrax broke out in the countryside around him, it was only natural that he would turn his microscope onto the blood of anthracitic sheep. And he discovered strange rod-shaped things existing there. A few other doctors-with-microscopes had already noticed "stick-shaped corpuscles" in the blood of anthrax victims. They looked like a new kind of bacilli. Often they would line up end-to-end in long winding chains, vaguely resembling a freight train. They might make one think of infusoria except these were found in mammalian blood instead of in water, and they didn't wiggle, or move, or grow, or change in any way that would show they were alive. The doctors-with-microscopes before Koch had even guessed that these sticks might have something to do with the disease. Maybe they were a consequence of the disease. Or maybe they were even the cause. But to most educated people of the time, that was a baseless and foolish guess. The French were still coming around to Pasteur's way of thinking, and elsewhere, most people still thought disease was due to internal factors. How could something so invisibly tiny cause enough harm to kill a huge cow? Those bacilli-that-look-like-freight-trains were probably just some kind of bodily residue, some debris of the disease. At worst, they were just harmless colonists seeking diseased blood as their natural habitat. They surely weren't the *cause*.

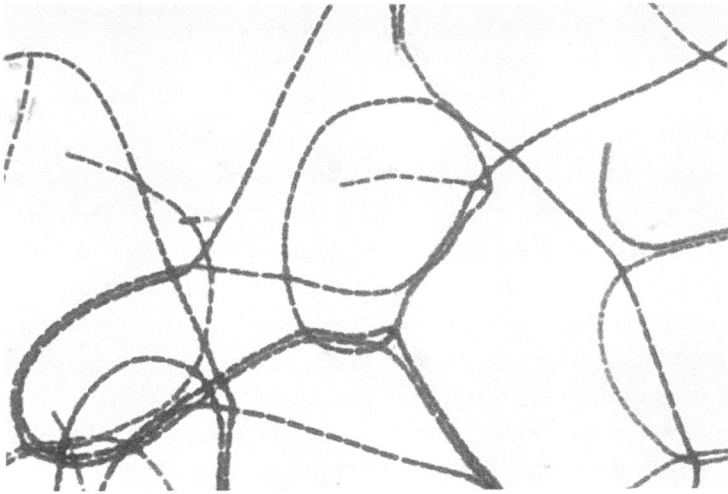

Anthrax Bacilli
Credit: CDC / Dr. James Feeley

Koch saw these sticks. He suspected that they might be the cause of the disease. And then he set himself to the challenge of *proving* it, one way or the other.

The Germ of Anthrax

If these new bacilli were the true "seeds" or "germs" of splenic fever, if their presence was the cause of the fever, blackness, and death, then these bacilli should always be present in the blood of diseased animals, and they should never be found in healthy animals. Koch began to gather his evidence by searching the blood of animals — hundreds of them — and passing sample, after sample, after sample, under his lens. In blood samples from diseased animals, as long as the disease had progressed sufficiently to show symptoms, he could usually find the strange trains of rods in the blood. In healthy animals, he never did. So far, so good. These bacilli definitely seemed to be connected to the disease somehow.

But there had been no evidence yet that these sticks were actually living things. Koch never saw them move, or grow, or change in any way. Maybe they lived a motionless, plant-like lifestyle. If so, that would make it more difficult to demonstrate that they were alive. But if those quiet sticks in the microscope field were really living parasitic germs, capable of growing and reproducing in their host, then Koch should at least try to show that they could multiply. Furthermore, if these bacilli were the true cause of anthrax, then they should turn a healthy animal into a sick one after they entered the animal for the first time. Koch needed to show that the change from a healthy animal to a sick one happened when there was one and only one other change: the entry of the bacilli into the animal's body.

72

But here Koch faced problems. Proving that the sticks were alive, and that they caused splenic fever, demanded experiments, preferably a great many of them. Putting together the rest of the puzzle and building the rest of the case against the germ was going to require *lab work*. But there was nobody to teach Koch how to do it. Except for Pasteur, nobody in the world knew anything about measuring or nurturing microbes. And even Pasteur had not yet tried to tackle human diseases. So, Koch single-handedly developed the art of handling diseases in a laboratory.

Lab Work

First of all, filling a laboratory with messy, noisy, hungry, diseased sheep was out of the question. Mice would be much better. But can mice be given anthrax? And how? Maybe one could poke a little diseased sheep blood into them? But then how could you be sure you weren't contaminating the mice with any *other* microbes from around the laboratory? (The hollow hypodermic needle had been invented a couple of decades earlier, but it is unclear if Koch knew of it or if he could obtain sufficient quantities cheaply.) With practice, Koch eventually worked out a reliable procedure for sterilizing wooden splinters, dipping them in anthracitic blood, and poking them into mice. The mice fell ill and died. Many mice, consistently. But had the mice actually died from anthrax or from something else? Simply being poked with a wooden splinter probably wasn't good for their health. Koch became an expert at murine autopsies. Sure enough, in the deceased mice, the blood and spleen were always black. What about the bacilli? Under the microscope went samples of blood and spleen. There were the sticks-in-the-blood.

Koch had invented a way to give anthrax to mice artificially in a laboratory by contaminating them with diseased blood. He had found a useful experimental animal, and he had shown that the disease can be transmitted by something living (or at least reproducing) in the blood. And he continued to find the mystery sticks in all diseased blood and never in healthy blood.

So the case for the prosecution was gaining strength. But the defense still had reasonable doubt. The bacilli had been linked to the disease, but they had not been shown to have any actual activity or measurable effect, nor even to be alive.

If these sticks-in-the-blood were living parasites, then perhaps it would be possible to nurture them and to grow crops of them outside of a host in their own "germ aquarium" of some kind. Could Koch find a way to *culture* the bacilli? Perhaps if he could do that, then he could watch them eat and grow and reproduce. He could watch them and handle them in their own test tubes or Petri dishes, rather than having to work inside a living sheep or mouse. He could observe and experiment with the parasites *in vitro*, rather than *in vivo*.

But how? If the bacilli are alive, what do they like to eat? Do you need to keep them warm? Koch reasoned that, if they thrive in mammalian bodies, they probably prefer to be kept warm, and they might like to live in mammalian fluids. But he also needed to keep them in a *clear* fluid, so he could watch them.

Anyone who has ever taught or participated in a middle-school science class may be familiar with cow eyeballs. They are readily available as a by-product of beef production and are relatively easy and fun to dissect. If you have ever dissected one, you may

74

remember the watery fluid that leaks out when you cut into the cornea. This is the *aqueous humor*, and Koch realized that it might be just the clear, nutritious incubation fluid he had been looking for. He jury-rigged a body-temperature incubator, he placed a small piece of diseased mouse spleen into cow eyeball juice, he placed this test sample into the incubator, and then he waited.

His incubator worked too well. If there were any anthrax-sticks in his culture, they were completely hidden by the jungle of other microbes that had overtaken his microbe-aquarium. Apparently, microbes had traveled in from harsher conditions, thrived in the luxury environment Koch had set up for them, and crowded out the anthrax-sticks. Now what to do? Koch needed to find a way to block out all other microbes from the surroundings, and culture only the anthrax-sticks by themselves.

Perhaps Koch could have used Pasteur's swan-neck flasks, but those wouldn't have made it very convenient to work with the microbes. Instead, Koch invented another very simple, ingenious solution. He placed a tiny shred of mouse spleen in a droplet of aqueous humor onto a thin glass slide, and then he placed the slide upside down over another piece of glass with a little hollow well formed into it. The droplet hung there, stuck to the ceiling in its own closed chamber between two pieces of glass, cut off from the outside world and out of the reach of dust or any other microbes.

This time, after incubation, the jungle contained only one species: the anthrax bacilli. After only a few hours, they had grown from a few threads clinging on to the bits of spleen, into massive tangles of twine surrounding them. They still didn't appear to wiggle or breathe or grow, but in those few hours, they must have multiplied because now there were many where once there were few. They had multiplied like a living thing. Koch had known they must, because a few of them on a wooden splinter could turn into a mouse-full a day later. But now he had clearly demonstrated the existence of reproduction outside the mouse under controlled laboratory conditions for the first time.

Would this new generation of bacilli cause anthrax? Koch extracted a small sample, and placed it, not into a mouse, but into another hanging-drop culture-well. He incubated this, and then repeated the process several more times, starting each new culture from a carefully chosen sample of the previous one. He wanted to be sure that he had a pure-bred culture of anthrax bacilli, with no trace of contamination from anything else. Finally, after eight generations, he found himself with a purified culture of anthrax bacilli that hadn't seen living animal flesh since their ancient ancestors, and with these far descendants, he inoculated a mouse. It fell ill and died. Koch confirmed by autopsy that it had black blood and a swollen spleen, and he confirmed by microscope that the blackened blood was filled with new bacilli.

The case for the prosecution was now pretty convincing. These tiny bacilli-that-look-like-freight-trains existed in all mice killed by anthrax, and they were never found in healthy mice. They could reproduce and be cultured, like living things, and their introduction into a healthy mouse was followed by death by anthrax. For the first time in history, Koch had made certain that one particular kind of microbe was the direct cause of one particular kind of disease.

What should Koch do next? He was a country doctor, and an entirely self-trained and self-made scientist, working alone. He didn't announce every new experiment, and he didn't participate in the usual back-and-forth of other scientists. He preferred to be rigorous and systematic, and to be sure of his convictions before he entered public debate.

Sometimes, having to answer the objections of others can help you make your own arguments better. But sometimes they just get in the way and waste your time. If you are a clear-thinking and hard-working researcher (or lawyer), it might be more efficient to think of objections *yourself*, and to answer as many of them as you can. Before you open your mouth, you anticipate possible responses, and try to answer as many as you can before other people can even ask them. You take your time, and you play *devil's advocate* for the opposition. You try to defeat your own arguments. This virtue of thoroughness and thinking ahead can be found in many of the heroes of our story, including Leeuwenhoek (who was also something of a loner) and Spallanzani. But Koch exhibited this virtue in spades.

So far, he had only demonstrated the relationship in his laboratory mice. But mice might be different from sheep. Now that Koch had reliable and well-practiced procedures, he could try larger animals. He methodically repeated his disease-and-germ tests with guinea pigs, rabbits, and eventually sheep. No matter the animal, the bacilli and the disease of anthrax always went together, as cause and effect.

Now, were anthrax bacilli the true cause of the anthrax disease? Almost surely. But one credible rebuttal still remained: the cursed fields. And Koch would not be content until he found the answer.

The Mystery of the Cursed Fields

We can suppose that the living germs of anthrax, like most living parasites, leave their host at some point and find a new one. Perhaps we could explain the cursed fields by supposing they are full of anthrax bacilli, deposited there by their previous host and lying in wait for the next host. But Koch's germs, if they were living things, could hardly be immortal or invulnerable. Like any living thing, they must die without food or in hostile conditions. In his laboratory, Koch had observed that his cultures weakened and disappeared when neglected too long or when kept in the wrong conditions. If the natural environment of these anthrax bacilli was the warm, rich, comfortable blood of mammals, then how could one explain the existence of germs in a mountain field which had been free of livestock for years and had been frozen and buried in snow and subjected to droughts and who knows what else in the meantime? What kept the tiny parasites alive in hostile fields season after season when they died on glass in two days?

Koch found the answer, and the final piece of the puzzle, in one of his failed, dried-up cultures. In the samples of faded and apparently dead germs, he noticed that the remains of the culture had filled with little oval speckles. Was this contamination by some other interloping microbe? Closer inspection revealed that the speckles were actually *inside* the corpses of the former germs. Was it some sort of decomposition? Was it some other kind of microbe that ate the bacilli? Koch kept that sample for a month and noticed no change. Then he decided to see what would happen if he put fresh cow eyeball juice on these speckles. Would they dissolve and go away? Would they grow and reproduce into clouds of speckles? No, they turned back into long chains of anthrax rods! It looked like the speckles were just a different form of the anthrax bacilli. Could these reborn bacilli cause anthrax, like the original pre-speckle bacilli? Yes, they could. Koch performed more experiments, and found that these speckles could endure and wait out cold, they could survive heat (up to a point), they could survive drying out, and they could remain infectious for months under a range of adverse conditions. The anthrax bacilli had a trick up their sleeves, and Koch's thoroughness had just found them out. Whenever times were lean or conditions were harsh, the bacilli could package themselves for storage. They could turn into these hard, durable, inactive, hibernating speckles, and the speckles could unpack themselves back into living bacilli whenever they were returned to healthy (for the bacilli) conditions. It was as if each bacillus could turn itself into its own seed, and then back again, whenever it needed to. Koch had discovered that some bacteria form *spores*.

So much for the mystery of the cursed fields. They were not cursed, they were full of spores of anthrax bacilli.

The Public Indictment

The year was 1876. Koch still couldn't name the mechanism by which the tiny germ caused damage. He couldn't explain *how* the life habits of the bacilli made spleens swell or blood turn black. But Koch finally felt confident enough to present his case to the public. He knew with certainty that these bacilli were the cause of the disease anthrax. He packed his bags with samples and equipment, rode a train to the prestigious University of Breslau, and delivered a three-day lecture, complete with demonstrations, illustrations, and plenty of microscopes. He presented his case thoroughly and methodically, with overwhelming evidence and unassailable logic. He won over the audience and the scientific community at large. The doctor and self-made scientist had expertly proven a cause-and-effect relationship. For the first time in history, Koch had securely linked a specific infectious disease to a specific microbial cause — the parasitic bacterium now known as *Bacillus anthracis*.

The Germs of Human Diseases

Anthrax is mostly a plague upon livestock, on herbivorous mammals and maybe rodents. Human beings can catch it from animals, but it doesn't usually spread from person to person, nor cause epidemics among populations. It is mostly a concern of individuals who work with livestock or those who travel to strange lands.

If you lived in Koch's day, you were probably far more worried about the chronic, spitting, bloody cough that was slowly consuming people from the inside out across much of Europe. "Consumption" was the leading cause of death at the time. Depending on the decade and the nation, it was responsible for one in every eight deaths, one in six, sometimes one in three. This is the disease that killed Emily Brontë, Frederic Chopin, and Doc Holliday. The only consolation, if you could call it that, was that the damage took hold and progressed fairly slowly, allowing people to remain functional for up to five years or so.

But what caused it? Nobody who became afflicted with the disease ever had any idea where or when or how their affliction began. (We now know that symptoms can appear months or years after infection.) Was consumption due to some flaw in nutrition? Was it an inherited disease? Nobody knew.

Sometimes doctors would examine the lungs of people killed by consumption, and they would find small lumps or "tubercles." In 1834, the disease began to be known by a new name: *tuberculosis*. In 1869, a French scientist caused tuberculosis to appear in rabbits by injecting them with matter taken from diseased humans or cattle, thus showing for the first time that tuberculosis is actually transmissible and contagious. It is more than a flaw in nutrition. There is an external agent. If *anthrax* is a contagion caused by a living parasitic microbe spreading from host to host and infecting one after another ... maybe *tuberculosis* is, too? Following his success with anthrax, Robert Koch marshaled his newly invented methods, techniques, and skills, and he redirected them at this devastation upon humanity.

Tuberculosis

Again we begin. If the disease was caused by a germ, then one should be able to find that germ in every diseased animal, but one should never find it in a healthy animal. Koch managed to obtain the corpse of a man recently killed by a severe attack of tuberculosis, so he could search for a culprit microbe. But this time, there was no germ to be found. The anthrax bacilli had been fairly easy to stumble across. These tuberculosis germs, if they were there at all, were well hidden.

Anyone who has ever played with a hobby microscope and searched for critters in pond-water knows that the world in the window is often not very colorful. Sometimes all you see is a glowing pool of light, with faint shimmery edges of shadows here and there. It can be like looking at a jellyfish or a clear glass bead underwater. Watery things swimming in water are much easier to see if you can *paint* them somehow. I'm not sure if Koch can be said to have invented the art of microscopic staining, but he was certainly one of the pioneers. In the previous few years, he had taught himself how to stain a variety of microbes with a variety of dyes. In the search for the culprit of consumption, Koch tried every stain and dye and combination he could, and he finally found a dye that revealed a new germ. He found a strange, artificially colored trespasser in the lungs of the dead man. This time, instead of chains of rods that look like freight trains, he found wispy threads, tinted faintly blue by his dye.

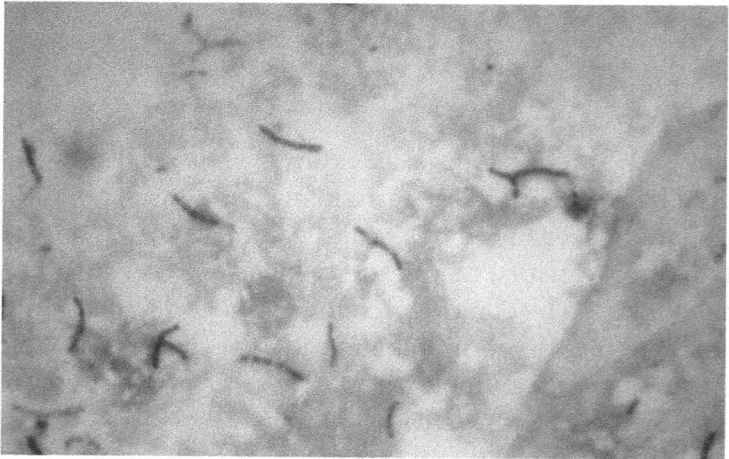

Tuberculosis Bacilli in a Tissue Sample
(Stained using the Ziehl-Neelsen method, which evolved from Koch's method)
Credit: CDC / Dr. George P. Kubica

As before, the hunt with the microscope now had to be followed by extensive laboratory testing. Could he give guinea pigs tuberculosis? He tried injecting them with ground-up tubercles. They died. Had they died of tuberculosis? Guinea pig postmortems revealed new tubercles inside the corpses. Now for the key question: Had wispy threads appeared in the new victims? The microscope revealed new generations of threads thriving in the deceased guinea pigs.

Koch was nothing if not thorough. He begged diseased tissue from hospitals and morgues across Berlin, and then he injected it into hundreds more guinea pigs, as well as mice, rats, rabbits, cats, dogs, chickens, pigeons, and a pair of groundhogs. He never found the wispy threads in healthy animals, only in animals that died.

Koch had his suspect. He had a solid candidate for the germ of tuberculosis. As before, now Koch needed to isolate, culture, and purify this suspect microbe, and then inoculate an animal with these pure-bred cultures, to see if this created tuberculosis in the animal. But again, this germ proved more truculent than the anthrax bacilli. This one did not like aqueous humor, nor would it grow in any other culture media that Koch tried. Maybe these fussy eaters needed *live food*? Or at least fresh blood? Koch figured out a way to make jelly out of sterile blood serum, and he tried growing his suspected thread-germs on that. At first, it looked disappointingly like another failure. But maybe this microbe was just an unusually slow-grower? After all, tuberculosis can take years to kill a person. Koch decided to wait and see. After two weeks, he finally saw spots. Tiny colonies of something had grown on the blood-jelly. What were they? Were these spots little cities of the suspect threads, or were they something else? The microscope showed that they contained the wispy blue-tinted threads. Koch had figured out how to care for his second microbe. He could grow this one in test tubes on blood-serum-jelly. His microbe husbandry expertise had grown to two species.

Having found a suitable home for them (with microbes, food and home are much the same thing) Koch set out to isolate, to purify, and to prepare pure cultures of one and only one microbe. He collected samples from dozens of his deceased animals and set about growing crop after crop, generation after generation, each time starting one culture from an extract of the previous one. After months of purification, he was ready to try the crucial test: Would an inoculation of this pure-bred microbe result in the appearance of tuberculosis? As before, one animal was not enough. In another attack of thoroughness, Koch injected his purified microbe-threads into guinea pigs, rabbits, hens, rats, mice, and monkeys, as well as animals not known to suffer from tuberculosis naturally: tortoises, sparrows, five frogs, three eels, and a goldfish. Animals in the naturally immune group never got sick, and the wispy threads never appeared in their blood. But the guinea pigs succumbed. (It is unclear what happened to the other mammals. Different animals have different levels of susceptibility to human tuberculosis, and I suspect Koch found erratic results with the others.) The guinea pigs had contracted tuberculosis, and the only change that had preceded that was the entry into their bodies of one kind, and only one kind, of microbe. Koch had satisfied another of his requirements for proof. Injecting a healthy susceptible animal with a pure-bred germ culture caused it to become ill with that germ's specific disease. The final requirement — finding a new generation of germs in the formerly healthy animal's body — quickly followed.

The microscope revealed that the next generation of germs had filled the bodies of the newly diseased rodents. The disease and death of the guinea pigs had been accompanied by the growth and reproduction of the living parasitic germs in their bodies.

Koch was now certain that these wispy threads were the true cause of tuberculosis. But he decided to see if he could fill in one more piece of the puzzle before presenting his final case to the jury of the world's doctors and scientists. As with anthrax, he wanted to understand the means of transmission. How do the tuberculosis parasites move from one host to another? All of his laboratory animals had been given tuberculosis by injection, but tuberculosis is primarily a disease of the lungs. People probably caught the disease normally, outside the laboratory, by inhaling it. Could Koch give his animals tuberculosis, not with a needle, but through the air? But how in the world can you spray deadly germs into the air safely? Koch's inventiveness again solved the problem. He built his animals an air-tight kennel with a spray nozzle, and then sprayed a mist of the deadly threads into the air. Without ever touching the animals with a needle, and with himself safely outside the chamber, he succeeded in killing them with nothing more than an airborne mist.

In 1882, Koch, the country doctor and independent scientist, traveled to Berlin and addressed a meeting of the biggest names in German science. The next day, headlines around the world trumpeted the news. The deadliest assassin of humanity had been found. Doctors flocked to Koch to learn how to culture the germ, how to make it appear under the microscope, and how to inoculate animals with it. As with anthrax, there was still no cure, but at least the dread disease could now be effectively diagnosed, and intelligent measures could be taken to avoid it and keep it from spreading.

Robert Koch had caught the culprit of consumption — the parasitic bacterium now known as *Mycobacterium tuberculosis*. His thorough efforts to answer all doubts and sort out all difficulties had securely proven a cause-and-effect relationship a second time. He had now logically, persuasively, and permanently identified the microbial cause of two major diseases.

Cholera

In the 1800s, epidemic diarrhea or "Asiatic cholera" simmered quietly in India, with occasional sorties into neighboring lands. In 1883, outbreaks erupted in Egypt and threatened Europe. Pasteur by this time had been spurred into action by Koch's discovery of the anthrax bacillus and had shifted his full-time energies into discovering and destroying the causes of disease. When cholera erupted in Alexandria, both Germany and France sent a germ-hunting team to Egypt to find the cause. Pasteur sent a couple of his assistants, and Robert Koch went himself with one of his own assistants. It was an odd sort of competition, inflamed by Franco-Prussian rivalry, but also strengthened I think by a distant respect between two giants fighting ultimately for the same thing. When one of the French team — a youthful newcomer named Thuillier — succumbed to the deadly disease he had been studying, the German team paid their respects at the funeral. Robert Koch helped to carry the casket.

Eventually, Koch and his dyes revealed a tiny new bacteria in the cholera patients. This one was not a stick-shaped *bacillus* or a spherical *coccus*, but a wiggly, comma-shaped microbe, a vibrating *vibrio*.

86

The Cholera Vibrio
Credit: CDC / Dr. William A. Clark

With a suspect identified, Koch then traveled to India, where the disease was abundant, to attempt to prove that this was truly the germ of cholera. (Animals do not suffer from human cholera, and if you can't work with lab animals, the only thing you can do is to go somewhere where there are many sick people.)

By now, Koch was an old hand at being a germ detective, and the wiggly cholera vibrio offered few of the challenges of the spore-forming anthrax germ, or the hard-to-find and hard-to-grow tuberculosis germ. Koch marched through his steps of proof fairly quickly this time. He searched every victim he could find, as well as hundreds of healthy people. He always found the vibrio in diseased patients, and never in healthy people, nor in any of the animals he searched, from mice to elephants. The germ was easy to grow and purify on standard beef-broth jelly.

The only weakness in his case this time was that Koch could not demonstrate that injection of pure cultures caused the disease, because he could not deliberately inject people.

Further investigations showed that the cholera vibrio is easily killed by drying out. Unlike the anthrax bacillus, it does not have the ability to transform itself into temporary weatherproof spores. However, unlike many parasites, it does not necessarily require a host. It can live just fine in water outside a body …as long as it can find water that sufficiently resembles the contents of a person's intestines. Koch showed that the water tanks in many huts were harboring the germ, and that it could sometimes be transferred by soiled linen. For the parasite, this ability to live in foul water meant that it didn't have to be too aggressive about finding a new host right away. But for human civilization, it meant that cholera, unlike the spore-forming anthrax or the airborne tuberculosis, could easily be stopped by basic sanitation. Modern water and sewage treatment systems have virtually exterminated cholera in developed countries.

For a third time, with rigor and thoroughness, with order and method, Koch had proven a cause-and-effect relationship — this time between the disease of cholera, and the parasitic bacterium we now call *Vibrio cholerae*.

Koch's Rules of Proof

Louis Pasteur was inspired (or perhaps provoked) by Koch's discoveries to return his attentions to the war against human diseases, and Pasteur would go on to become a giant in the business of combating disease. Pasteur would eventually give humanity the means to wipe away many infectious diseases, and he saved millions of lives. Koch never provided a successful cure for anything. But Koch was king when it came to rigorous proof. Koch's conviction of those first three germs of disease wiped away confusion and debate, and allowed future disease-fighters to concentrate their fire where it belonged — on the microscopic parasites that cause the harm.

But constructing certainty from scratch is like assembling a three-dimensional puzzle. Even with centuries of observations and experimental data to work from, earning a conviction in the court of science is still a lengthy and difficult process. Consider how careful and methodical Koch had to be to prove his case: he always started by identifying a suspect germ, an unusual microbe in a diseased patient. He then showed that this sus-pect microbe is always present in diseased animals and never in healthy animals, and he did this in as many examples, across as wide a range of conditions, as possible. He then found a way to isolate and purify this microbe and to grow purebred crops of it on some kind of organic food in a test tube. This confirms that it is a living thing, and enables the next step. Perhaps the most convincing step in the proof is to prepare a pure injection of that microbe, and only that particular microbe, and show that the inoculation is followed by the appearance of the dis-ease. The effect follows the cause. If the only thing that changed was the introduction of a particular microbe, one can be pretty certain that that microbe was the physical cause of the subse-quent disease. Finally, Koch would re-isolate and re-identify the microbe from the blood or the tissues of the sickened animal. This shows that the microbe has reproduced and filled the body as the body became sick. Koch didn't yet know how microbe anatomy worked, or the precise way in which the germs caused the damage, but at least he could show that their reproduction corresponded to the progression of the disease.

Koch always followed this methodical, rigorous procedure, and he always ended with conviction. If you are searching for truth, if you want your conclusions to be sound, if you want your quest for knowledge to end in certainty, this is what you must do.

Later in life, Koch settled into government advisory work, and published his rules of germ-hunting in a formal way in a government report. These rules of proof later became known as "Koch's postulates." As Galileo may have inherited some ideas from Renaissance thinkers before him, perhaps Koch inherited some ideas from Francis Bacon or John Stuart Mill. Koch's Postulates certainly bear a resemblance to "Mill's Methods" for identifying a cause-and-effect relationship. But like Galileo, Koch built his own legacy. He was responsible for the triumph of the germ theory. He was the first to *prove* a cause-and-effect relationship between specific parasitic microbes and specific infectious diseases.

THE LEGACY OF THE GERM-HUNTERS

Galileo was an intellectual fountainhead in 17th century Italy, pouring out a river of inspiration and creativity that ran in streams through all of subsequent history. In the field of disease, I think it can be said that Pasteur and Koch were both similar fountainheads, in 19th century France and Germany, respectively. They both built world-famous institutes around new fields of medicine and trained a generation of disease-fighters, and many of these disciples would go on to make significant new discoveries on their own.

Pasteur had no training as a doctor, and he wisely hired two medical assistants to help him through most of his disease research: Charles Chamberland and Pierre Paul Émile von Roux. Under Pasteur's guidance, this team discovered that weakened germs can induce immunity without causing disease, and they built on that discovery to engineer the first three artificial vaccines: those for anthrax, chicken cholera, and rabies. To this day, the Pasteur Institute is world-renowned in the treatment and prevention of diseases. The number of lives saved by Pasteur's vaccines, and by all of the subsequent vaccines patterned after them, cannot be counted.

In 1884, Chamberland invented a new kind of water filter, one with pores so small it could remove all microbes. One could take water from this filter and examine it with the highest-power microscopes and find no life. Chamberland wanted something that could supply their laboratory with pure sterile water, but eventually people started using his device to filter other things — germ cultures, blood serum, and such. In the 1890s, people were searching for the unidentified germs of new diseases, and they discovered that with some diseases, there was an invisible germ. You could filter the extracts from diseased plants or animals and confirm with the microscope that these extracts contained no microbes, and yet somehow the clean fluid still made healthy animals sick. Was there a poison instead of a germ? No. Whatever it was, it could kill animals generation after generation, as strong after ten generations as after the first. So this "poison" must reproduce and replenish itself in each new animal. It must be something living. Was it some kind of living poisonous fluid? Or was it a germ, but a germ so small that it could pass through the filter and that couldn't be seen even under the microscope? If so, it would have to be a germ far, far smaller than any known bacteria. Whatever it was, it became known as a *filterable virus* — using "virus" in the original sense of "unhealthy filth or slime; poison." Later the word "filterable" was dropped, and the name "virus" was officially applied to this entirely new category of invisible germ. Chamberland's filter had enabled the discovery of viruses, and figuring out exactly what viruses are would occupy much of the 20th century.

In 1888, Roux tried filtering cultures of diphtheria, thus removing the bacteria that were known to cause the disease. He discovered that the filtered culture still made animals sick, even without the bacteria. But this time, the sick animal could not infect any other animals, and no diphtheria bacteria filled the animal's body. The animal had been sickened, not by the bacteria directly, but by the *toxin* they had manufactured. The animal had been poisoned ... *by the bacteria*. Before Roux made this discovery, nobody knew for sure *how* the lives and habits of parasitic germs harmed your inner functions and made you sick. Roux showed that in at least some cases, they do so by making poison. The germs don't attack or eat our tissues directly, they poison us with their products. As fermentation bacteria make alcohol or lactic acid, so some germs make lethal toxins.

Like Pasteur, Robert Koch also had colleagues, trainees, and disciples. One of the most famous of these was Paul Ehrlich. Ehrlich was a wizard with dyes. He figured out how to stain various things *inside a living body*, and he discovered for the first time that there are different kinds of white blood cells. He was also fascinated with the problem of how to kill a living parasite without harming the host. How do you create a drug that will attack only the things at which it is aimed, and nothing else? Ehrlich searched far and wide for these "magic bullets," and he eventually found one. Paul Ehrlich created the first antibacterial drug, or *antibiotic* — a syphilis treatment called Salvarsan.

As the 20th century dawned, so did the age of modern medicine. Many diseases caused by the microbial "whales" (i.e., the protozoa) would soon be wiped out by sanitation. Other protozoal diseases, such as malaria and sleeping sickness, would be stopped by pest control. (Near the turn of the 20th century, it

was discovered that many microbial parasites, especially protozoa, hitch a ride from host to host in mosquitoes or flies.) To
fight bacterial infections, we now have a spectrum of antibiotics,
although overuse has been producing more and more resistant
strains of bacteria. Broken limbs can be healed, and women can
give birth in modern hospitals with negligible risk of death from
hospital diseases. Pasteur's dream of making parasitic maladies
disappear from the face of the globe was beginning to come true.

At the same time, however, a few perplexing puzzles remained.
For one thing, not all germs were easy to find. Some were very
elusive, especially those invisible "viruses." For another thing,
we had discovered that germs cause disease, but we had also
discovered that microbes aren't rare attackers, scattered and
hidden in nature. Microbes are *everywhere*. Instead of asking
why we get sick, the new question was: why *don't* we get sick?
If we live our lives in an ocean of microbes, why aren't we sick
all the time? What keeps us *immune* to most of their effects? The
solutions to these two puzzles would make fascinating stories
of their own ...

* * *

Knowledge is not like an apple that you can pick from a tree.
Knowledge is like a castle that you have to build. To learn that
germs exist, what they look like, how they live, and what we
can do about them was not quick or accidental. Like any other
scientific achievement — the discovery of the structure of the
solar system, or the nature of atoms, or the nature of DNA —
the discovery of germs required far more than a few chance
observations and a few guesses. It required the invention of new

instruments, and the development of new skills. It required extensive experimentation, with careful comparisons and controls. It required perseverance against opposition. It required curiosity, initiative, diligence, inventiveness, and attention to detail. Above all, it required thorough thinking. Achievement requires virtue. The discovery of germs required the accumulated virtue of dozens of hard-working scientists spanning centuries.

While doing my own research in preparation for writing this little book, I gathered information and took notes from many sources. One of them was the classic book *Microbe Hunters*, by Paul de Kruif. If you'd like to read more about the discovery of germs, I recommend starting there. I'd also like to conclude with a quote from that book. To describe Robert Koch's achievement of finally discovering the cause of an epidemic disease for the first time, de Kruif wrote:

> So it was that in these three days at Breslau this Koch put a sword Excalibur into the hands of men, with which to begin the fight against their enemies the microbes, their fight against lurking death; so it was that he began to change the whole business of doctors from a foolish hocus-pocus with pills and leeches into an intelligent fight where science instead of superstition was the weapon.

This is what careful observation, relentless testing, and thorough thinking can do.

www.ingramcontent.com/pod-product-compliance
Lightning Source LLC
Chambersburg PA
CBHW071438210326
41597CB00020B/3847